How to do Your Own
WOOD FINISHING

How to do Your Own WOOD FINISHING

By Jackson Hand

POPULAR SCIENCE

HARPER & ROW
New York, Evanston, San Francisco, London

To Charlie Feeney

Library of Congress Catalog Card Number: 67-10842
ISBN: 0-06-011778-8

First Edition, 1966
 Eight Printings

Second Edition, Revised and Updated, 1976
Fourteenth Printing, 1982

Manufactured in the United States of America

CONTENTS

INTRODUCTION

FINISHING—and particularly refinishing—articles made of wood is becoming an increasingly popular activity. Its greatest attraction lies in the way a new and beautiful finish increases the value and the usefulness of a piece of furniture. Many is the item of junk, forgotten in a corner of the attic, that has been reborn into new and useful beauty under the brushes and patience of a refinisher. Many is the undistinguished piece of unfinished furniture that has blossomed into stardom in a youngster's bedroom. Many is the product of a home workshop that is transformed from a good job into a masterpiece by the right finish, carefully applied.

Wood finishing, simplicity itself, required a combination of tools everyone knows and can handle, techniques anyone can master, materials anyone can afford. The techniques and tools we use today are very much the same as those used by the John Goddards, William Saverys and Duncan Phyfes of the past. But some of the materials have changed tremendously—for example, varnish. There is no comparison between the varnishes we have today and those of only ten or fifteen years ago. Not only are they now tougher, but they form a clearer film, they handle better, and they do a great deal of your work for you.

A modern satin-finish varnish produces a surface very close to that obtained by meticulous rubbing. It turns smooth and satiny as it dries. This doesn't put an end to the old "piano finish" that comes from several carefully rubbed and smoothed coats of varnish; you still want it in some situations. But it is nice to be able to brush a varnish on some such semi-critical area as a paneled wall, and have it dry to a rubbed finish.

But, you may say, we have had satin-finish varnishes for a long time. There's a difference, though. The old nonglossy varnishes contained solids which broke up the surface, disrupting the gloss. These solids were semi-opaque, and they gave the varnish a certain degree of cloudiness. Today the semigloss is achieved chemically, and the film is quite translucent.

You will find that all the finishing materials mentioned in this book are on the shelves of your favorite paint store, with the exception of a few specialist items you may have to order from such craftsman supply houses

1

as Albert Constantine & Son, 2050 Eastchester Road, New York, N.Y. 10461, and Craftsman Wood Service Co., 2727 S. Mary Street, Chicago, 60608. Anytime a technique is recommended which may involve the sort of materials not normally carried in regular paint stores, you'll find mention of the fact, so you won't be caught short.

In many instances throughout the book specific brands are mentioned or shown in photographs. The reason why some names are given is that we know no better way to designate a *type* of product than to name specifically some examples. It would not be possible, of course, to mention *all* the brands which will provide a given result, and rarely will you find that the brand names mentioned are the only products that will perform the function involved. Your dealer should be able to tell you which of the products he handles is parallel in performance to one this book might mention.

The literature on wood finishing is filled with much misinformation, cant, and old wives' tales, as well as pet formulas and rituals. You'll find nothing of that sort here. We're interested in the *best finishes for wood,* and in the research conducted in preparation for this book our sources were most often the technical directors of finish manufacturers. We took their advice into a typical home workshop and put it to test on many different woods, in many different situations, by many different methods. As a result, you'll find here the things that the chemists of finishing *know* to be true—*proved* to be true in the hands of the guy next door. And if he can do it, you can do it.

Jackson Hand

CHAPTER ONE

HOW TO SELECT
THE RIGHT FINISH

MOST of us who appreciate the beauty of wood—its colors, lusters, textures—are actually admirers of wood *finishes*. When we see wood that is completely natural, and perhaps rub an approving hand across its surface, we are envisaging the wood that will be when the finish is on it. A walnut plank, planed smooth, is good to look at; sanded carefully and treated to a penetrating finish it becomes beautiful, filled with fires and depth it could never show in the raw. The most elegant mahogany crotch veneer, gracing the top of a Georgian piecrust table, is dull and low-key until the varnish hits it. A piece of No. 2 pine, undistinguished in the pile at the lumber yard, gains its own, forthright dignity when the proper stain highlights the grain pattern and emphasizes the honesty of its knots.

The wood *and* the right finish—that's where the beauty comes from. Picking the right finish, and it must be right in all respects, is the first step.

For any project, there are usually two or three finishes that are about equally appropriate. Choosing the one you consider best is a matter of balancing several factors against each other. In general, these are the considerations:

BEAUTY. In wood finishing, beauty is more in the eye of the beholder than in almost any other homecraft. You must consider *wood texture*—or lack of it. Some finishes are perfectly smooth. Others leave the pores of the wood open. You must consider *color*. Stained or unstained, every species of wood has its characteristic hue—and this hue changes from one finishing material to another. For example, shellac gives cherry a color quite different from that it gets from a coat of varnish. You must consider *sheen*. A good varnish, laid on in a full coat and blessed with the good fortune of dustlessness, is as smooth as the smoothest glass. It can be rubbed to any degree of sheen. Or—you can shortcut the long road to a satiny rubbed finish by using a varnish that dries to a semiluster—or even dead flat.

As you work more and more with finishing materials and techniques, you'll discover the look you like best. If you will study carefully the finishes you see, analyzing what you like about them, you'll learn what to seek after. When you know that, you'll have no trouble finding the color, the sheen, the texture which combine to give you exactly what you want.

DURABILITY. Some projects demand the most abuse-tolerant finish you can find. The need on a cocktail table for resistance to water, alkali, acids, and other chemicals leads you straight to a polyurethane varnish or other extra-quality varnish. A rosewood jewelbox, on the other hand, might deserve the beauty—and the trouble—of a French polish, a finish not intended for rough usage but durable enough to last several lifetimes resting on a dressing table.

Many experienced finishers believe that you never need to sacrifice good looks in exchange for a rugged finish. They cite the high abrasion resistance of urethane varnishes which are clean and clear, permitting all the beauty

Which finish goes where? The Georgian desk surely deserves a fine rubbed finish. The whatnot shelf above it does too — but there is too much fretwork, so it gets satin varnish. The mirror frame is a strong candidate for French polish. Below it is a decorated Hitchcock that should be unchanged, or perhaps varnished to protect decorations. The endtable needs a urethane varnish to protect its mahogany top. Penetrating resin finish is for the carved walnut wood parts of the sofa. In the foreground, the oak stool needs nothing. The plant stand takes penetrating finish on the old chestnut base and top, with an overcoat of urethane varnish to protect the top.

of the wood to show through. And they cite the penetrating resin sealers which are considered to be the ultimate in good looks for open-pore woods and are tough enough to be leaders in floor finishes.

EASE OF APPLICATION. This consideration gives a wood finisher more pause for thought than either good looks or durability. Some finishes are so quick and easy a child can do them excellently. Others take so much time and patience that they barely justify themselves on any other basis than purism and tradition.

Again, the nature of the project may affect your decision. The nice little piece of unpainted furniture you buy for a child's room may serve all practical and decorative needs with a couple of quick coats of shellac, to make it easy to dust and clean. A piece of modern walnut might thank you for your kindness in giving it the easy-as-pie brush-on-wipe-off penetrating finish. You might not be satisfied, yourself, however, until you have put the final coat of wax on a mahogany ogee mirror that was really worth the many coats of varnish, the slow drying, the meticulous rubbing.

But halfway through this time-consuming, blister-raising job of rubbing, you'd be entitled to look longingly at the can of satin-finish varnish on your shelf. It is *almost* as beautiful as rubbed varnish. If you do decide to hit the final coat of satin-finish with No. 400 wet sandpaper or with pumice in oil, then wax it, even an expert would have trouble pinning the short cut on you.

To some wood finishers, the ritual and the routine are part of the fun. Others would like to get the job done. Both will find the methods and materials for their preference as they read on.

THE CHOICE OF MATERIALS. As you know, a finish is a combination of material and a method. There are differences—some slight, some enormous —between the *way you handle* the various materials, as well as in the way they perform for you. Each major finishing material has its own chapter, later in this book. Now, however, is a good time to highlight each of them, as a means of helping you to make that basic decision: *which.*

Varnish. This is the most popular clear finish for wood, because it is fairly easy to use, provides good durability, and gives a color well liked by most people. Its durable, clear coating is often selected for furniture, floors, trim, and all other forms of natural wood finishes, including brightwork on boats. It is made by mixing tough resins in oil-derived vehicles, along with dryers and other chemicals which provide such characteristics as semi-gloss drying when it is desired. When the vehicle evaporates, a film of resins remains, stuck tight to the wood. In most cases this film is dust-free in an hour to four hours, ready for handling in about eight hours, and dry enough for sanding and recoat in about twenty-four hours. Some special varnishes are made of resins which are hard in about eight hours, but do not reach full "cure" for perhaps a week.

The color varnish gives to wood is basically the result of oils it contains. It intensifies wood tones, tending to darken dark areas proportionately more than light areas, thus increasing the pattern of wood grain. The general effect is to give the wood warmth—a reddishness. You can get a quick close approximation of the color by brushing turpentine or linseed oil on a sample of the wood.

Some varnishes have a slight yellow color, the natural hue of the resins. There may be a tendency for this yellow hue to deepen as the finish ages. Older varnishes turned almost orange with time; new resins, particularly the urethanes, are lighter and go yellow very slightly. By and large, you can forget yellowing with these materials.

"Synthetic" varnishes, some of them containing acrylic and vinyl resins, are extremely clear and do not change color at all. Their only shortcoming is a tenderness as far as abrasion is concerned.

The durability of varnish is excellent—unsurpassed in some respects. Urethane and polyurethane varnishes have the highest resistance to abrasion and to ordinary chemicals. Those based on other resins, such as alkyds and phenolics, have slightly less resistance to abrasion and to chemicals. They tend, however, to be brittle. For this reason they scratch "white," and may show a fine network of cracks, in time, particularly if the wood is subjected to wide changes in humidity. The wood swells and shrinks in a degree greater than the varnish film can handle. In extreme cases the varnish actually chips off.

Varnish is widely used outdoors. A kind called "spar" is specifically intended for exposure to the weather. However, the action of the sun plus other rigors of heat and moisture usually causes any varnish to fail outdoors in six months to two years. When you consider a clear finish for use on the outside of the house, for garden furniture and the like, you must expect a refinishing job fairly often. For that reason, most people turn to pigmented exterior stains, even for wood like cedar, cypress, and redwood —or to enamel.

Varnish is among the more difficult materials to apply well, although it goes on slapdash with great ease when the work is uncritical. A good varnish brushes easily and levels well. It will hold a good coat on vertical surfaces without sagging. But due to slow drying, it is extremely difficult to avoid dust marks, unless you can work in an area that is entirely dust-free. There are techniques for overcoming dust marks—but they represent some extra work.

You can buy good varnish in spray cans—and you can use it in regular spray guns. Results are excellent in both cases—subject however to the same problems with dust.

Shellac. This is probably the second most popular clear finish for wood, owing to its quick-drying properties, low cost, and ease of use. Wood finishers often use shellac almost any place you'd use varnish, although it is not as water-resistant. Some of the most beautiful finishes are done with shellac. The vehicle in shellac is denatured alcohol, and the resin is a natural gum secreted by an insect called the lac bug, found in India and neighboring countries. The gum, cleaned and refined, is dissolved in alcohol. In use, you dilute shellac even further with alcohol, producing a thin liquid which brushes easily, flows out smooth, dries quickly.

Shellac comes in two colors: white and orange. The latter has a deep amber cast. On the wood, however, there is very little difference, and most finishers use the white in all cases. On the other hand, there are experts who prefer the orange on dark woods, the white on light-colored species.

Shellac gives wood a warmth and fire similar to that produced by varnish. Only personal experimentation will tell you which you like best. On woods such as cherry, mahogany, walnut, teak, rosewood, and other dark species, the color is particularly good.

No great claims for durability can be made for shellac, in situations of hard usage. It is smooth enough and flexible enough to withstand normal wear, and it does not scratch white the way a brittle varnish does. It resists the tendency to check, craze, and flake off. On projects which are not likely to suffer abuse from water, chemicals, strong detergents, and particularly alcohol, you can expect a lot of life from shellac. Even on floors, it gives excellent performance, particularly when it is reinforced with a good wax job. But—if a primary consideration is toughness there are better materials.

One of the major reasons for the popularity of shellac is the easy way it goes on. Thinned with alcohol to a consistency not much thicker than water, it flows out and leaves brush strokes with virtually no edges. Quick to dry, it lets you knock off the high spots with 6/0 or finer sandpaper and recoat in an hour or so. Thus, the multicoat requirement of shellac is often offset entirely by the ease and speed with which each coat goes on.

Shellac, like varnish, comes in spray cans. Although this form presents some convenience, it takes away your ability to thin shellac to your liking for good built-up finishes. And, with the spray can, you lose the economic advantage of shellac's low cost.

Lacquer, the third of the major clear finishes, involves still another type of resins—usually cellulose nitrate in combination with other manmade materials. However, some of the newer resins are used in the most modern lacquers, including acrylic, vinyl, and others. Owing to a rather loose lexicography in the paints and finishing industry, lacquer and varnish seem to overlap somewhat, and an acrylic clear may be thought of as a varnish by one manufacturer and as a lacquer by another. This is further complicated by the growing use of petroleum derivatives in finishing materials which formerly contained acetates. Modern lacquers have vehicles which are highly sophisticated blends of acetates, alcohols, and hydrocarbons, and through the control of resins and vehicle components, lacquers can be made slow-drying (to be brushed on) or fast-drying (for spraying only). When you shop, keep in mind that you can put on a brushing lacquer by brush *or* spray, but you cannot handle a spraying lacquer with a brush. It sets up too quickly.

Lacquers are usually water-clear, and their effect on the color of the wood is much like that of water. They intensify grain and color, but do not materially change the hue. There is a special variety of brushing lacquer called "lightener" which leaves the color of the wood virtually unchanged. Use it whenever you want the final finish to retain the color of the original wood.

The ability of lacquer to withstand abuse is good, as indicated by its wide use on floors, under the trade name "Fabulon," in itself an excellent brushable lacquer for finishes other than floors. The lacquer film is hard, difficult to scratch. But it has low flexibility and usually shows a tendency to craze after a year or two. Most furniture factories spray on a lacquer finish, the fastest method for mass production.

Brushing lacquers, used thin and in multiple coats, are easy to use, comparable to shellac. They rub well, and you can build any thickness of film with good adhesion between coats.

PENETRATING FINISHES. Finally, your choice may be a penetrating finish—one that is entirely in the wood, never on it. The in-the-wood finishes grow in popularity every day, not only because they are so easy to use, but because they give wood a fantastic amount of protection without obscuring its beauty in any way. Penetrating finishes are little used on trim, but are excellent on floors, on furniture, paneling, cabinets, and other interior jobs. In general, as the separate chapter on these materials will explain, they are best suited to open-pore woods, and do not perform ideally on woods such as cherry, maple, and other fine-grained species.

SOME WOODS DEMAND CERTAIN FINISHES. One of the most important considerations in selecting a finish is the wood itself. Some species—mahogany, for instance—can be finished almost any way—stained, bleached, filled, natural, in-the-wood. Others don't seem right unless they have one specific finish. A good example is teak, which in the eyes of most finishers would be profaned by anything other than a penetrating, in-the-wood finish which reveals the texture and the grain. The rule of thumb is that fine-grain woods are best with finishes that build on the surface. Maple, birch, pine, cherry, gum, basswood, beech, and similar species may be sealed with a penetrating finish, for good adhesion, but their topcoats should be varnish, shellac, or lacquer rubbed, glossy, or semiglossy.

Open-pore woods such as oak, mahogany, chestnut, pecan, walnut, ash, and the like look right with a penetrating finish or one that is on the surface. Here the choice depends on the nature of the project. Traditional mahogany was nearly always done with a filled, smooth, rubbed finish; for that reason, traditionally styled pieces you finish or refinish should be smooth. With modern mahogany, open-pore finishes may be more appropriate. This is true with walnut too. However, the absolute opposite approach is used with oak. Traditional oak was always open-pore finished and much of it had no finish at all. Oak used in modern design is filled more often than not—sometimes with a colored filler which makes a novelty of the wood pores.

OPAQUE FINISH. The fact that we have touched only on transparent finishes thus far in this chapter does not suggest that the only appropriate finishes for wood are those which let the wood show. Far more wood is painted than varnished, shellacked, or lacquered. Some of this painting is quite utilitarian; you paint a small step-stool to give it some protection and perhaps to make it match the kitchen woodwork. Other painting has pure beauty as its objective, and opaque finishes can be very lovely.

Your materials in the opaque finishes parallel those in the clears. Enamel is nothing more than good varnish with pigments in it. In the same way, colors are added to shellac and to lacquers.

Colored shellac is intended as a primer and sealer for wood which is to be painted or enameled. It comes in black and white; the white can be tinted with universal colorants. While it is an excellent material to use as the first coat on big jobs, it has very little value in small-scale work. Generally, the job is better and the nuisance level lower if you use an enamel all the way.

Insofar as the average person is concerned, colored lacquer means lacquer in spray cans. *Brushing* lacquer is not easy to find in stores, although you can buy an excellent quality of *spraying* lacquer in almost any conceivable color at auto-supply outlets. The advantages of lacquers in opaque finishes exactly parallel those in clears: they are quick to dry, and multicoat built-up finish is possible in a day's time.

When speed is not an overpowering consideration, high-grade enamels generally produce a longer-lasting finish and one that is somewhat easier to apply. Used in the same way as varnish, in several coats with a smooth-out sanding between coats, a good enamel builds to a perfectly level finish which rubs beautifully and can be waxed to produce a result every bit as lovely as "Chinese lacquer." The durability of a good gloss enamel exceeds that of a good varnish, indoors or out, because the pigments provide protection as well as color.

Enamels come glossy, semiglossy, and flat. The flat would be "paint" in the lexicon of most finishers, because the pigments are rather coarse (this produces the flatness) and you'd rather put them on a plastered wall than on wood. Durability of enamels decreases from glossy to semiglossy to flat.

SYNTHETIC AND LATEX FINISHES. Not long ago an expert in the finishes industry predicted that a time would come soon when all finishing materials would be of the latex type, thinned with water. Several resins—or latex materials—are used for both clear and colored finishes. These include the well-known acrylic and vinyl latices which appeared first in paints, then in enamels, and finally in clears. They have excellent flexibility and resistance to mars and normal abrasion. They cut easily, however, which means they do not stand up well under harsh abrasion.

If you choose one of these materials, because of its rapid drying or the ease of cleaning brushes, etc., with water, don't expect it to build into a high-grade rubbed finish. The latex finishes do have their place, however, in routine varnishing and enameling.

CHAPTER TWO

BASIC FINISHING DECISIONS YOU MUST MAKE

No TWO finishing or refinishing jobs are exactly alike. Differences in the nature of the item, its value, the condition of the old finish, the amount of repair work involved—all these and many more personal conditions make it necessary to analyze and appraise the job carefully before you begin. The kind of finish, the *degree* of finish—even whether to finish at all—may be determined on the basis of this appraisal. In addition, sizing up the job helps you determine the materials needed. Finally, it helps you plan your work so that you don't find yourself halfway through the first coat of varnish when you run across a loose spindle you should have taken care of earlier.

Since the details of finishing are usually established at the outset when you build a project from scratch, most appraisal involves *refinishing*. This includes old floors as well as old furniture.

THE OLD FINISH. The first question is, *"Must the old finish come off?"* This is important, because any time you don't have to strip old paint or varnish, you save a great deal of work. Most of the time, you can put a new opaque finish (enamel, glazed, etc.) over an old one, and it is not rare to work on a piece which needs nothing more than a rejuvenating coat of varnish to restore it to beauty and usefulness.

Check these signs of deterioration of an old finish:

1. *Is the old finish adhering well?* Examine it carefully for signs of flaking. Look for cracks in a large alligator pattern. These indicate a brittle finish which doesn't come and go with the wood as it changes dimensions with changes in moisture content. It is usually unsafe to overcoat such a surface, since it may transmit its cracks to the new finish under continuing conditions of expansion and contraction. One good way to test adhesion is to make a serious effort to scratch the finish with the edge of a quarter or half dollar. If you find that, with hard pressure, the edge does not powder or flake the finish off clear to the wood, you can be pretty sure that the old finish is on to stay. Make this test in a spot which doesn't show conspicuously, because you are almost sure to abrade the surface somewhat—

Checking and white spots on this old finish are signs of bad adhesion. When a putty knife takes it off right down to the wood, you know the finish must come off. Anything you put over it is sure to fail because the old finish will fail, even if resins were to penetrate the fissures and make them less evident.

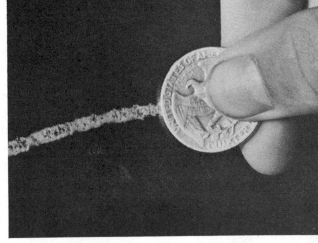

One accepted test of brittleness and adhesion is to make a rather serious attempt to scratch an old finish with the worn edge of a quarter. Make the test in an inconspicuous place, because it will probably whiten as shown here. But, if you can't scrape through to the wood with relative ease, you probably can get away with rejuvenation instead of removal.

even produce a glaring white scratch on old varnish. This scratch will all but disappear when you put new varnish over it, but it might show in some place where it is easy to see.

2. *Is it smooth?* Many old finishes craze and alligator so markedly that you could never put a level finish over them.

3. *Is it chipped?* Look closely at corners and edges, as well as much-used areas such as drawer fronts, doors, and tops. If the finish has chipped off down to the bare wood or what appears to be the primer or sealer, you may be facing two problems. The chipped condition suggests poor adhesion and a brittle finish. In cases of minor chipping, you may sand the edges until they blend smoothly into the wood. Then prime or seal the bare wood, perhaps using two coats so you can be sure that subsequent materials won't sink in. With luck, the patch will never show. *However, your chances of doing this successfully with a varnish finish are virtually nil.* Before you can rub the chip out, then rub out the patching, you have

This photomacrograph shows the alligatoring typical of old built-up finishes. The varnish has turned quite opaque. An idea of its thickness shows at the edge of the bare wood (light area.)

spent more time than a strip-and-refinish job would take. However, it is often possible to refinish only the top—or a drawer front—or a similar independent area without redoing the entire piece. Just take pains not to spill remover on places where you want to retain the old finish. The biggest problem in this operation is matching the colors—a problem covered in the chapter on staining.

Whenever you find signs of poor adhesion, roughness due to crazing, weathering, etc., or much chipping, you'll save trouble in the long run if you remove the old finish. In the presence of any or all of these deficiencies, paint removers usually work their fastest and most efficiently, and what you *should* do doesn't turn out to be hard to do. The following chapter discusses the entire process of finish removal.

CLEANING THE OLD FINISH. Although a finish may be sound enough to recoat, chances are pretty good that it is too dirty to take a new finish without the risk of poor adhesion. Of course, you must never attempt to recoat a glossy surface. It is even less wise to put a finish over a surface which may be dirty or oily, or coated with invisible chemicals with which the new material will not be compatible. Over the years, even with careful maintenance, a film of dirt and wax and grease tends to build up on furniture, paneling, trim, and the like. In fact, *many a wood finisher has discovered that, once this coating of dirt is removed, all the piece really needs is rewaxing and it looks as good as new.*

But—clean means meticulously clean, if you intend to recoat an old finish. If a project is fairly rugged and can stand the dowsing, scrub it with detergent. Fill a pail part way with warm water and pour in a good lacing of any good household detergent. Or mix a half a cup of trisodium phosphate (paint stores sell it) in about two quarters of water. Go to work with a scrub brush, dipping it in the pail often, to give the cleanser plenty of chance. When you have scrubbed it clean, rinse the piece with a garden hose. As the water sluices off, watch for any areas where it balls

Many times an old finish is merely dirty, and returns to usefulness with a good scrubbing. Don't be afraid to use detergent. If the old finish is damaged by the scrubbing, it should be renewed anyway. Scrubbing is also good technique before you recoat an old piece, to make it clean enough so the new finish will adhere well.

You can save work many times by stripping old finishes off parts of a piece, rejuvenating the rest. Here a table top is stripped clean. Drawer fronts, legs, aprons, etc., will get light sanding plus a clean-up with a multisolvent varnish cleaner. Top gets a regular finishing schedule; the other parts need only one coat of varnish.

up as water does over wax or oil. Go at these spots with stronger detergent or even with scouring powder. You want the piece clean—absolutely clean. Rinse again, and let the project dry overnight.

Another way to clean an old finish perfectly for recoating is with special high-solvency chemicals paint stores sell, such as Wil-Bond or Liquid Sandpaper. These products are particularly good on delicate projects which might be harmed by soap and water. Used carefully, these chemicals will not ruin delicate finishing details you might want to retain. Used vigorously, they will clean an old finish even to the point of removing gloss, giving you a perfect surface for recoating. Many finishers clean old finishes with turpentine, alcohol, lacquer thinner, naptha, and similar chemicals, since these are among the ingredients you find in the commercial solutions.

It is impossible to overemphasize the importance of removing gloss on an old finish if you expect the new one to stay on. This is particularly true of glossy enamels and oil paints, which may develop a surface coating composed of oil and a type of calcium powder. Nothing will stick to this surface film. It must be washed off with detergents— and since it is usually quite invisible, you can be sure *only* if you wash the piece. Once the washed and rinsed glossy enamel is dry, take away its gloss with Wil-Bond or Liquid Sandpaper—or with actual sandpaper used in a fine or very fine grit.

RETAIN DECORATIVE DETAILS. Many old finishes include antique paint and decorating effects to which a piece may be almost entirely indebted for its charm and value. Many early American pieces carried painted finishes in black, gold, ochre, green ochre, and other earth tones, on which proud owners or local artists painted decorations that ranged from clusters of cherries to full-rigged frigates. Naturally, today such decoration is worth far more than the utility of the piece as furniture, and many finishers are willing to devote more time and patience to maintaining these examples of early Americana than they would expend merely bringing the wood beneath into an acceptable finish.

To retain such decoration and make it permanent, you simply coat it with a new transparent finish—usually shellac or varnish. Start experimentally in an inconscpicuous corner, working on a tiny area, and see if you can risk cleaning the surface, first with water, and then with soap and water, then with a chemical cleaner. As with any other old finish, the cleaner you can get it, the better, but you must be cautious. Some chemicals may take the decoration off entirely. When it is clean, check it with a small spot of varnish or shellac. If the test spot dries hard and clear, feel safe to go over the entire piece. Since you would not dare rub such an overcoating, for fear of cutting through the decoration, you might want to use a polyurethane semigloss varnish.

The decorations on the back of an old chair and other pieces give them much of their worthiness as pieces of furniture. Don't lose them. Delicate cleaning and careful coating with flat varnish will preserve the artwork.

For restoring large-scale work, such as floors, interior trim, etc.. paint stores sell a group of products known as renovators. Their action is partially cleaning, partially solvent, and partially coating. Usually, they must be applied with a rather severe abrasive action, such as that of a mechanical steel-wool buffer. In effect, renovators remove dull and opaque surface dirt and wax. They tend to fill minute cracks in the film. They provide a thin overcoat for the old finish, and provide a perfect base for another finish coat if it is needed. Although these products are mainly intended for restoring or renovating floors without the need for sanding and refinishing, they are workable on smaller projects, using hand-applied steel wool, preparing the old finish for a new topcoat.

REPAIRS AND REFINISHING. It is not only the condition of the finish which you must study in appraising a refinishing job, but also the mechanical condition of the piece. Quite often, when a great deal of patching or replacement is needed to put a project into shape, you may decide in favor of an appropriate opaque finish instead of a varnish, shellac, or lacquer. Although there are excellent patching materials (see the chapter on minor repairs), an absolute match in color between them and wood is not easy. Good match between old wood and new can be equally difficult; if you must replace important and conspicuous parts, give some thought to paint, unless an opaque finish would be entirely out of keeping with the style and character of the piece.

CHAPTER THREE

HOW TO GET
THE OLD FINISH OFF

WHEN you must remove old paint or varnish before you put on a new finish, you have a choice of three methods.

1. *Paint and varnish remover.* In liquid or semipaste form, these are combinations of various chemicals which soften the old finish so that it can be lifted off with gentle scraping or washed off with water. Contrary to some belief, there is no difference between a paint remover and a varnish remover, and the industry tends to refer to the material simply as "paint remover."

2. *Mechanical removal.* For most of us, this means sanding with belt or disk or (in case of floors) drum sanders. Sometimes finishes are removed mechanically by sandblasting, but this is a process hardly suited to most home projects.

3. *Heat.* High, scorching temperatures destroy the film and make it easy to scrape off an old finish.

Of the three, chemical removal is infinitely superior, leaving the other two methods impractical except for special circumstances. An electric-element heater is a good device for removing paint from a house, and it can be used on other, smaller-scale work when surfaces are flat. However, the amount of residue it leaves on the surface is enough to call for chemical means to clean it up. This suggests the idea of using heat as a rough remover when there are so many coats of paint (not varnish or shellac) that cutting through them would be slow and costly with chemical removers. *Never let anyone work at paint removal on your property with an open flame.* The danger of fire is altogether too great.

The trouble with sanding, which is fairly quick, is that you cannot take off the finish without also taking off some—or much—of the wood beneath it. This is all the more true with a power sander. You might be willing to remove some of the wood on a rough project, but generally speaking any fine work cannot tolerate removal of the surface of the wood. A major reason for this is that wood that is at all old has taken on a surface coloration (even under finishes) and if any of this color is removed it must all be removed, to avoid unpleasant blotchiness.

The coarse, open-coat sandpapers normally used for paint removal leave the wood quite rough. This often creates hours of work, smoothing the surface again.

The true place of sanding in paint and varnish removal is in clean-up following the use of chemical removers.

HOW TO PICK A PAINT REMOVER. At any paint store, you can buy a quart of paint remover for about $2—or you can spend about $4. Why the difference? The price is established, generally, by the cost of the chemicals used. Most experienced refinishers believe that the costlier removers are the real bargain, because they are the most efficient. The least costly removers will soften paint so that you can scrape it off. Usually, however, their formulas include paraffin, which coats the project and must be meticulously removed with some such solvent as turpentine or paint thinner. Otherwise, the subsequent coat will not dry. There are waxes of a different type in the costlier paint removers—and in some of them the wax does not remain as residue on the wood. Removers formulated this way are often labeled "no clean-up." However, they do not leave the wood entirely clean; you may still find a residue on it which must be sanded away to expose absolutely bare wood. In the upper price ranges, you find "water-wash" or "washaway" on the can, and this means that the formula includes emulsifiers which let chemicals and waxes mix with the water and rinse away. When you do the job properly with these removers, the wood is absolutely clean and bare.

There are other changes in paint removers as you move up the price scale. For one thing, the fire hazard is relatively great in the cheaper kinds, since the chemicals are highly volatile. In more costly formulas, methylene chloride is introduced, and when it is present in sufficient quantity, it acts as a damper on other chemicals which burn, making the material nonflammable. Methylene chloride is a high-solvency ingredient, too, so it does double duty in a paint remover. There are also differences in toxicity and the danger of breathing fumes. Such possible hazards must be labeled on the cans, in conformity with government regulations.

The major advantage of the more expensive materials is the speed, the thoroughness, and the simplicity of water wash-up. Unless you are working on a project involving delicate veneers with old glues which water might loosen, it is pretty hard to justify the use of anything less handy than water-wash—or at least no clean-up—removers. All the important producers of paint removers have fairly complete lines, in all price ranges.

When you must apply remover to a vertical surface, you'll appreciate the semipaste kind, which holds a better coating without sagging. As you work with them, however, keep in mind that they may become more liquid as the old finish softens; be prepared for runoff.

HOW TO USE PAINT REMOVER. Removing an old finish is at best a messy job—and if you approach it with a willingness to be messy, you'll

Whenever possible, work on a horizontal surface, turning the project to bring successive sides up. Best way to get a full layer of remover on is to pour it, except on curved or vertical surfaces.

Spread the remover in as thick a layer as possible, or flow it by tilting the piece from side to side, corner to corner. Do not brush out. You need a good quantity of material to work on the finish. Avoid thin spots; you'll just have to coat them again.

After about twenty to thirty minutes rub a finger in a circular pattern in the remover. If your finger — not your nail — cuts through to the wood, the finish is soft enough for the next step. Some people may have skin sensitive to removers and should wear rubber gloves for comfort.

Gentle abrasion with medium steel wool often speeds the action of the remover toward the end of the softening process. Do not apply much pressure or you'll squeegee the remover off and end its action entirely.

Final step is plenty of water and swabbing with fine steel wool, a wad of toweling, or similar soft material. Use plenty of water unless flooding might soften delicate veneer glues. Note newspapers on compartment shelves to keep remover from splattering on them; their finish did not need to be removed.

get it over with most quickly. You'll do a better job, too. Put on old clothes and rubber gloves. Spread papers around the working area. Then, save your talent for fine work until later on, when the time comes to put the finish back on.

The basic rule for using paint remover is to use lots. You will understand why this is important if you consider what the paint remover must do. It must lie on the surface in sufficient quantity and for a sufficient time to soften the old finish. Its enemy is evaporation. That is why there are waxes in formulas. They tend to coat the solvents and prevent them from vaporizing. But you must put enough of the stuff on to cut through the finish before it evaporates in spite of the protection the wax gives. As a psychological stimulus to the use of *plenty,* buy plenty in the first place. Don't start off on a project with a pint of remover and find yourself skimping because you should have bought a quart.

These are the steps to a good clean job with any remover:

1. Spread it on—or pour it on thick. Don't brush back over the remover once it is laid out. This would be tantamount to stirring it, causing faster evaporation. "Lard" it on with the flat of the brush. The less coverage you get, the better.

2. If the material seems to dull over or get skimpy in spots, brush on more. Keep it wet, all over the surface you are treating.

3. In about half an hour, maybe less, the thick coating you put on has done its work. Test the degree of film softening by pressing your finger into the surface with a small circular motion. If, by finger pressure only, you can penetrate to bare wood, the remover has done its work. (Don't scrape; if you let the remover soften the film until finger pressure will cut through it, you'll never have to scrape.)

4. If at the end of half an hour you cannot rub your finger through to bare wood, the reason why is that the remover evaporated before it could cut through all the old finish. Now you can choose between two procedures, equally effective. Either brush on another application of remover over the half-cut sludge, or slough off whatever comes easily with

19

a scraper or other means. Then apply more. The experts say it is lazier to put on more remover over the untouched sludge, but it is better in the long run to squeegee off what comes easily, then let the new application of remover go on from there. After twenty minutes to half an hour, all that is left, anyway, is emulsifier, thickener, and other nonsolvent chemicals. They don't harm the effectiveness of the new material; they do not help it either. Removing them at a halfway stage makes the job less messy.

Why wash-away is the easy way. When you use a wash-away remover this is what you do: Put it on as thick as it will lay out on a flat surface or suspend itself on a vertical surface. Recoat any areas which go skimpy. Test with a finger rub at twenty minutes. If your finger cuts through, turn to toweling or fine steel wool and water to rub loose the sludge. You can insure a complete softening and loosening by gentle abrasion with medium steel wool. Flush away the residue with plenty of water and a bristle brush. Use a soft wire brush to work the old finish out of carvings, crevices, fluting, reeding, etc. Wipe off as much water as you can with toweling or rags. Let the wood dry overnight. Then, take care of sanding and other required preparation of the surface.

Nothing beats the garden hose and a big wad of steel wool for quick and complete cleansing of the wood. Unless you are working on delicate veneers, plenty of water, following plenty of wash-away remover is the quick and easy way to a bare surface.

Many refinishers prefer to work outdoors, which is a good idea, *if you keep out of the sun.* Temperature is no factor in efficiency of methylene chloride removers; they work as well cool as warm. However, warmth increases the speed of evaporation. The stuff boils at 110-degrees F. Work outdoors to avoid unpleasant or harmful vapors, if you like—unless it is warmer outdoors than it is in the garage or the basement, with ventilation. Even in freezing temperatures, methylene chloride takes off varnish as readily as at 68 proper degrees.

IF YOU'RE A BIG-SCALE OPERATOR. Professional refinishers like this technique: Fill a big tank with a wash-away type of remover. The tank should be big enough to dunk a chair or a drawer—or a chest of drawers. To keep the stuff from evaporating, pour an inch-thick layer of pure water on the top. The methylene chloride is heavier than water; it stays down, and the water prevents loss due to evaporation. Put the item for stripping in the tank, leave it there for half an hour, take it out and jet-spray it with water. It will be absolutely clean. Since the solvents do not use themselves up, this setup needs only replenishment of volume through carry-off.

ANSWERS FOR SPECIAL PROBLEMS
- If you must remove one of the modern finishes— catalytic coatings, epoxies, urethanes, etc., standard removers may not have kept pace with

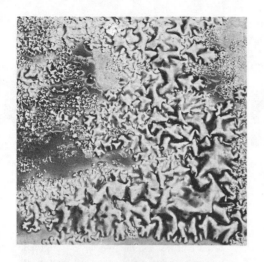

Action of the remover on enamels is different from that on varnish. Opaque finishes usually wrinkle up, layer by layer. Dark unwrinkled areas here need reapplication, since the remover is no longer cutting enamel.

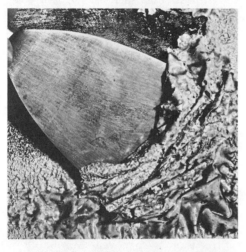

With more remover added to problem areas, the entire film is loosened to the wood. This type of wrinkling off is typical in situations where an enamel has been brushed on over old varnish that was not properly scarified or cleaned. Adhesion is poor.

Paint sinks into cracks around spindles and other joinery, and needs extra soaking with remover. Pour a puddle into a chair seat, as an example, then flood it into the cracks as part of the process of cleaning up the seat.

Convenient way to handle legs of furniture is to set the leg in a small can partly filled with remover, then brush upward on the legs. Excess drains back into the cans.

them. An electric-element remover will usually dishearten the toughest finish of this sort.

● When there is stain under the finish, it may or may not lift off with the film. The same is true of filler in coarse-grained woods such as walnut or mahogany. Water-wash removers take away more stain and filler than any others, but if there is stain left, try a quick wash with Clorox or a standard wood bleach such as Blanchit. Oxalic acid in a strong aqueous solution usually kills analine stains. Keep in mind that such bleaches also lighten the hue of the wood while they are removing stain.

● Furniture made in the past fifteen or twenty years probably is water stained, and there may be some "blending" to make the color of the wood uniform. Avoid wash-away removers and excessive sanding *if you want to retain this blending.* Many refinishers prefer the come-what-may color of natural wood and deliberately remove this blending with soapy water and bleaches. Very fine sandpaper completes the water-stain removal.

● Since there is no sense removing finish from areas which do not need to be refinished, use masking or newspapers—or care—to keep remover off such surfaces as drawer sides and interiors, the insides of cabinets, the backs of doors, etc. If you smear these places with remover, you'll mar them to a degree which may demand complete removal.

● If you use a putty knife or other scraper, be careful not to dig into the wood, which is sure to be softened at the surface by the remover. Although this softness disappears as the remover dries, it makes the wood delicate while it is saturated.

When you work on flat surfaces, it's quick and easy to use a broad knife to scrape softened finish over the edge into a tin can. Be careful with the knife; removers soften wood and make it easier to scratch.

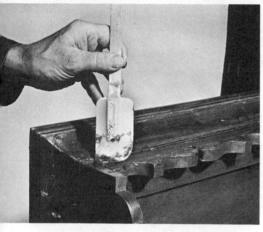

Rubber kitchen scraper makes a good squeegee for concave and convex surfaces as well as curves when you work with removers which are not formulated for water washaway.

Much the same advantage you get from waterwash removers comes from the use of a detergent and warm water to scrub off the "nowash" variety. Detergent replaces the emulsifiers you get in the water-wash removers.

Save yourself work by keeping remover off any surface which doesn't need to be stripped. Prevent accidental spills by plugging screw holes for pulls, etc.

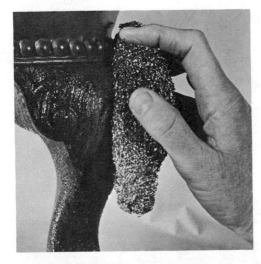

Uneven surfaces such as carvings, ball-and-claw feet, and the like are easy to scrape, using a "Chore Boy" kitchen abrasive pad. Follow this with plenty of rags or with water.

Get into the intricate carvings with a medium-stiff wire brush. Be careful not to use too much pressure, or the carvings may be damaged. Often a brush full of remover on carvings after the rest of the piece is cleaned up softens finish in the cracks and speeds removal.

Broken end of a small stick has protruding fibers which act as a miniature brush to clean grooves.

Pour some paint remover in a bowl and dip pulls, hinges, and other hardware in it. Paint and varnish come off metal almost instantly. Clean away residue, then spray hardware with clear lacquer to keep it bright.

Electric heating elements soften and blister old finishes so that a putty knife lifts them off, although usefulness of such tools is limited mainly to flat surfaces. Paint reacts better to heat as a remover than clear finishes.

Example of the clean stripping you get using a water-wash remover by the methods suggested in this chapter. After the remover had done its work, water was used with fine steel wool.

● When only the tops of tables, chests, etc., need refinishing, work the remover up to a quarter-inch or so from the edges, to avoid dribbling it down the sides. Then, take off the remaining finish with sandpaper.

● On a big job, when cost is a factor, use a cheap remover to take off the heaviest coating. Then finish up with a quick application of wash-away.

● After a water wash-up has dried, if you find tiny areas where finish still remains, don't waste time reapplying remover. Sandpaper will usually cut quickly through the spots of finish, since they have been weakened by the remover.

● Don't be afraid to use warm water and a detergent for clean-up. It will not harm the wood except in the case of fine old veneers.

CHAPTER FOUR

HOW TO CLEAN UP
THE STRIPPED WOOD

EVEN the most meticulous work with paint remover leaves some residue on the surface and in the grain. There may be bits of finish which the remover missed. In cracks and crevices there are always deposits of dried remover. If you use an inexpensive remover, you may have a film of wax coating the wood. All of these things interfere with the smoothness of the new finish, possibly with its drying, and quite likely with its adhesion. They affect the way stains penetrate and color the wood. They must be removed so that you have perfectly clean wood to work with.

Usually, you must wait for the stripped project to dry thoroughly before you can see where the additional work is needed. Discoloration reveals waxes or thin coatings of dried remover. Quite often dried bubbles in cracks and carvings have the look of tiny volcanos. Examine the underside of tops, the inner surfaces of legs, the bottoms of stretchers, and other places which may have escaped careful wiping or rinsing.

Don't even trust smooth, conspicuous areas that look clean. Test their readiness for a new finish by sanding them lightly with a piece of fine flint sandpaper. If the paper clogs, you know the surface is still dirty. If it turns up nothing but good, honest wood dust, you know it is clean.

Fortunately, post-removal clean-up is usually quick and easy. Only rarely will you find it necessary to apply more chemical remover. Most of the remaining dirt, you'll find, is old remover that has dried hard. It re-

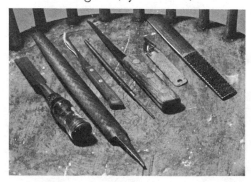

Typical collection of tools you need for cleaning the last traces of old finish from a stripped project are: chisel, half-round rasp, kitchen knife, rat-tail file, boning knife, beer can opener, four-in-hand file. Flecks of finish imbedded in the seat of chair are a type of difficult residual finish discussed in text.

27

Four-in-hand rasp has fine and coarse, half-round and flat, for excellent versatility in removing old finish. Special feature of this tool is no teeth on the edges, so you can avoid cutting where you don't want to.

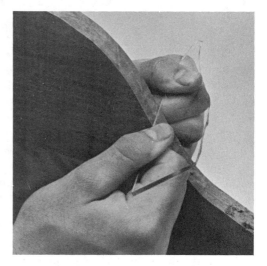

Edge of broken glass is so sharp it cuts end-grain smoothly, removing penetrated stain and other finishes. Always make the stroke downhill in relation to the grain. Change to fresh glass as it knicks and grows dull.

sponds quickly to sandpaper, to scraping, and other tricks. Now and then you'll find a small spot which, for some reason, has never been touched by remover. It is usually quick and easy to sand such spots clean, without using any more remover.

The more difficult clean-up is on end grain, particularly where end grain shows on turnings and curved coped pieces. These areas usually hold stain that has penetrated more deeply than in other areas. Two methods are commonly used to clean up these situations:

1. Use broken glass to scrape away a very thin layer of wood, never enough to alter shapes, but enough to remove the discoloration. Be careful working with glass not to cut yourself. As a safeguard, put little pieces of masking tape over the glass where you take hold of it. With a little care, however, you can handle glass without hurting yourself, and nothing beats the smoothness of the cut it makes. You can buy scrapers complete with handles and if you keep them sharp they work well. However, the myriad shapes and sizes and angles you get from broken glass make it the best scraping material.

2. Use bleaches. Oxalic acid usually lifts the colors of stains. So does ordinary household bleach, brushed on full strength. Sometimes you must combine scraping with bleaching. The action of the scraper removes pigments which are on the surface or in surface pores. The bleach removes the color of *aniline dyes* which may have penetrated so deeply that you would not want to remove enough wood to eradicate them. When you use bleaches to remove stain and discoloration, chances are they will lighten the adjoining clear wood somewhat too. Uniformity of color can usually be restored when you stain, however.

Another form of debris which clings to the wood after you take off the remover is specks of finish which may be lodged in nicks, crevices, cracks, and other depressions in the wood. Some of these depressions may be mars and scratches, some may be part of the design. Use ingenuity cleaning out these bits of old finish—and discretion. Work carefully in grooves and other elements of design so as to avoid spoiling the appearance of the piece. Sometimes the age and character of the project may suggest that you leave the flecks of old finish in place as a method of accenting antiquity. Also, you must always take into consideration the need for filling mars and scratches with some material which may not always tone the same color as the wood and therefore may not look much better than the residual old finish.

The ingenuity comes in the selection of tools. Anything in shop, kitchen, or garage counts, as long as it can be used to dig, scrape, pry, or gouge away material you want to get rid of. Most useful are rasps, files, awls, knives, and chisels.

A wad of steel wool in your hand, applied with a twisting motion, cleans up spindles and other rounds. Although this action is across the grain, steel wool does not cut sharply enough to cause problems.

Kitchen scouring pad is a good tool for cleaning fairly large grooves. Press it into the depression with your thumb, working with the grain. Follow with finer steel wool or medium to fine flint sandpaper.

Degree of bleaching you get from oxalic acid depends on concentration in water and temperature of water. Leg and part of stretcher here were lightened with about a tablespoon of acid powder in a cup of fairly hot water. Experiment to find the strength you like best.

A good test for the cleanness of a surface is to give it a few swipes with medium or fine flint paper. Clumping and filling such as that on lower edge of paper shown here indicate residual wax and other materials, which can usually be removed by re-wiping with lacquer thinner.

Although sanding as a step in preparing wood for finishing is covered in detail in the next chapter, sandpaper has its place in preliminary clean-up too. In medium and fine grades, it will remove dirt and discoloration better than any other mechanical method. Use flint paper, which is economical, rather than costlier garnet or aluminum oxide paper, since clean-up work fills the abrasive quickly; you use up a lot of paper.

Important: When you work on a piece that is old, remember that a great deal of its beauty and color, when refinished, will come from the natural aged look of the wood itself. *Do not over-sand, or you may remove this relatively thin surface layer of aged wood.* Countless hundreds of old pieces, lovely with the coloration that comes only from years, have been converted into ordinary run-of-the-mill furniture by thoughtless refinishers who were overzealous with sandpaper during clean-up. This warning applies with equal force to bleaching techniques.

The final step in cleaning up wood for refinishing is a thorough wiping with plenty of rags and a solvent. Many finishers use lacquer thinner. When there is a chance of wax, flood the surface with turpentine and wipe it while it is still wet. No chemical works as well as Cleanwoode and similar products paint dealers sell specifically for making wood absolutely clean.

CHAPTER FIVE

HOW TO MAKE
SANDING EASIER

No FINISH can ever be smoother than the wood it goes on, and that is why the experts say there is no such thing as a good finish over a poor sanding job. Sanding to *smooth* as opposed to sanding to *clean*, starts when the old paint or varnish has been removed or when a new project has reached the assembly stage. The transition between clean-up and sanding comes when you give the project a thorough cleansing with lacquer thinner, turpentine, or one of the commercial wood-cleaning fluids. This cleansing not only helps guarantee good performance of finishing material, but by removing gums, waxes, and dirt, it makes sanding easier. Sandpaper lasts longer without clogging, and it cuts the clean wood faster.

MECHANICAL SANDERS. Sanding takes time. It is hard work. Few people find it much fun. For that reason, mechanical sanders are very popular tools. If you are considering adding one to your list of wood finishing equipment, here are the types:

Vibrating sanders. These sanders have a vibrating pad driven by an electric motor which gives the pad either an *orbital* or a *straightline* motion. Smaller electric sanders have magnetic drives which produce only a straightline action. The circular motion of the orbitals makes it inevitable that half the strokes will be across the grain—which is a poor sanding technique. This crossgrain motion contributes to swirl marks on the wood which are visible no matter how fine the sandpaper may be. The swirls must be sanded away *with the grain,* to finish the job. That is why the orbital type of sander is not the best for wood finishing. There is, however, a sander which produces either orbital or straightline sanding at the flick of a lever. With one of these, you can take advantage of the cutting speed of orbital sanding, then smooth the job with straightline.

An advantage of this group is that they use regular sheet sandpaper, cut to fit. Thus, you have full range of sandpapers to choose from.

Belt sanders. With a belt sander, you get straightline sanding and great cutting speed. You can level off rough plane marks in a few seconds using a coarse belt, then switch to finer grit for smoothing. However, this same fast cutting can be a drawback when you work on refinishing projects

—particularly antiques—since it is easy to take away more wood than you mean to.

Belts are limited in grit and variety, compared with sheets. They seem costly when you buy them, but there is a great deal of sandpapering in a belt, particularly if you use it only on bare wood and keep it from clogging.

Disk sanders. The motor spins a rubber disk to which a circular piece of paper is fastened. Swirl marks are usually serious, and for that reason disk sanders aren't worth much in wood sanding. However, fitted with buffing or polishing pads, they are excellent for quick work on any polished finish.

Drum sanders. These big machines are intended for floor work only. They use heavy sheets of abrasive paper which wrap around the drum and lock tight. They must be accompanied by a smaller sander—usually disk type—since the drum sander can't get close to edges or into corners.

Although mechanical sanders are capable of doing the whole job in work that is not critical (floors, paneling, trim, etc.), you cannot expect them to do fine finished sanding. On most furniture you want a final smoothness that can be produced only by hand sanding after the machine has done the rough work.

HOW TO PICK THE RIGHT SANDPAPER. The right sandpaper and enough of it so you don't have to skimp on fresh, fast-cutting grit make sanding easier. For work on wood, there are three kinds of sandpaper grit:

Flint. This is the material with the backing that is like grocery store bags and the grit that actually looks like sand. Its only virtue is its cheapness. You use it in situations where a gummy, resinous, waxy surface quickly fills and clogs the paper; when it fills, you throw it away. It is hardly worth your time to monkey with flint paper on good, clean wood. It is not sharp enough. It cuts slowly. It dulls rapidly.

Garnet. This is the paper with the reddish grit. It is sharp, hard, clean working. The price is two and a half time that of flint. It will do five times as much work. Most careful and meticulous wood finishers are partial to garnet paper for their fine sanding.

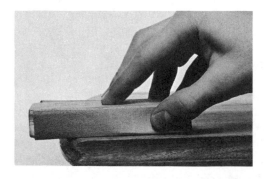

On any flat surface, always use sandpaper with a block. When you work to the edges, keep the pressure on the back of the block so it won't rock over and cut too deep.

Aluminum oxide. This grit is about equal in performance to garnet, although it wears a bit longer. More stores carry it, in grits up to 6/0, than carry garnet. It is a man-made abrasive, very sharp and fast cutting. More aluminum-oxide paper is used in wood finishing than any other kind, in sheets, belts, and disks.

You have some choice in paper backing too, although the manufacturers usually make your decisions for you by matching the grit and the backing properly. In principle, the coarse papers should have a heavy backing. Finer grades, intended for more delicate work, need a thinner, softer backing. On the back of the paper look for the letters A, C or D. The 3M papers are labeled "A wt." Behr-Manning uses the grit and backing designation together, as "120-C." In addition, papers on light backing are often called "Finishing," while papers on heavier paper are called "Cabinet." Generally, all flint paper, fine or coarse, is on the same backing—a grade of paper less durable than that used for garnet and aluminum-oxide grits.

For most projects, you need several sizes and shapes of blocks, to get into corners and confined places. Shown with the homemade felted blocks here are a metal and a rubber variety you can buy at stores.

Why should you be concerned with the backing? For hard sanding, the stiffer, tougher backing of the coarser grits—cabinet papers—lasts longer. For fine work, however, this same toughness tends to produce scratches. With the softer backing of the finishing papers you avoid this, and the greater flexibility lets you follow contours of shaped or coped surfaces more readily.

A third choice you have when you buy sandpaper is abrasive spacing. Among garnet and aluminum-oxide papers, you can buy open-coat or closed-coat. The reference is to the spacing of the grit on the paper. Open-coat has grit over as little as 50 percent of the paper—*and that is all you need.* There is more work and more life in the paper because of the open spaces

Don't tear sandpaper. If you do, the ragged edges often scratch fine work. Cut it by scoring it firmly on the back with an awl. Then pull the pieces apart.

which gradually fill up with sander dust, but which release the dust when you whack the paper over the corner of the workbench.

Open-coat papers are usually in grades 80 and finer. Coarser papers need to be more rugged and tend less toward clogging, since there is naturally more space between particles of abrasive.

Some special nonfilling papers are available (an example is Behr-Manning's "No-Fil" garnet) which have an extra-wide spacing between grains, plus a pattern of graining that gives a very smooth surface on wood without undue filling.

Flint papers are not designated open-coat or closed-coat.

The major decision in the selection of sandpaper is, of course, the grade. This means the fineness or coarseness of the grit, and there are three different methods of designating the grade. Flint paper is usually labeled "medium," "fine," "very fine," and so on. Some brands of aluminum oxide carry grit numbers, such as 80, 120, 220, etc. The higher the number, the finer the grit. Other aluminum oxide—and most garnet—papers are labeled by the grit number plus the old standard 3/0, 6/0, 8/0, and so on, with which most wood-workers are familiar. It makes no difference which method of designation you go by—as long as you get it more or less in your mind that 6/0, 220, and very fine are about the same—as are 3/0, 120, and fine. The chart on page 36 gives comparisons as closely as possible.

The 120 grit is about as fine a paper as you need for large, uncritical areas such as floors and paneling, and for more critical areas which are to be finished with paint.

For fine work, to be finished clear, or with fine enamel, or with a rubbed clear, 220 is essential. More demanding workers use sandpaper finer than 220. This means that you may have to heckle your dealer a little for a special order of 240, 280, 320, or 400 garnet, because most stores today carry a standard merchandising package which stops at 220. Constantine's catalog shows 240 garnet as the finest; Craftsman's Wood Service shows 280. The industry considers anything finer than 220 "industrial," not "retail."

To sand curved surfaces on which a block won't work, make a pad by interleafing two pieces of paper as shown here. Fold them grit out. As one working surface fills or wears out flop the pad — then reverse the interleaf to produce two new sides.

When you work on a project with many identical moldings, it's efficient to make a special felted sanding block to fit the exact curve.

Fold a small piece of paper twice to sand in small crevices and corners. As the paper wears right at the crease, change the fold.

Toward the coarse end, you reach a grit which may roughen more than it smooths. At about 80 grit—even 100—the only use for sandpaper is to *shape* the wood, or to remove a considerable thickness. When this is the need, sandpaper may be the proper thing to use, but chances are you need a rasp or a scraper or some other cutting tool. This is true, also, when you attack a very rough surface. Perhaps it should be planed or scraped before you turn to sandpaper.

SANDPAPER SELECTION CHART

USE	GRIT NUMBER	GRADE NUMBER	WORD DESCRIPTION	BACKING
Rough sanding and shaping	80	1/0	Medium	D
Preparatory sanding on hard woods	120	3/0	Fine	C
Preparatory sanding on soft woods	100 or 120	2/0 3/0	Fine	A
Finish sanding on hard woods	220 to 280	6/0 to 8/0	Very fine Extra fine	A
Finish sanding on soft woods	220	6/0	Very fine	A

BEST SANDING TECHNIQUES. Most of the special tricks and techniques in using sandpaper are demonstrated in the photographs that illustrate this chapter. There is, however, a basic routine followed by most craftsmen:

1. After you've completed the clean-up, start with medium paper—about 3/0—and go over the entire project. Work with a sanding block, and make every effort to sand uniformly. Some workers deliberately count strokes, in order to give all the wood the same amount of sanding. This attention is worthwhile at every stage as you make the wood smoother and smoother. The reason is that sanding tends to affect the degree in which wood accepts stains and other finishes. Make the experiment yourself; sand a scrap of wood, then sand half of it twice as thoroughly. Dust it and apply a stain. You'll see the difference. In some work, the variation may not make much difference. But, on the top of a maple table, for instance, you could give yourself a serious problem in color uniformity if you sanded much more in one area than another.

2. Switch to a finer grade and again sand the entire project. Some finishers like to use 4/0, then make another switch to 6/0. Others make the jump directly from 3/0 to 6/0. When you sand hardwoods like maple and birch and oak, the intermediate grade speeds the job, since the 4/0 knocks off the roughness 3/0 leaves, and makes it easier to reach final smoothness with 6/0. The hardness of the wood and the fineness of grain also dictate

Sanding edges, avoid rounding the corners by lifting from the back of interfolded sheets with your forefinger while pressing with your thumb. Balance of pressure vs. lift makes paper cut flat across the edge, leaving corners unrounded.

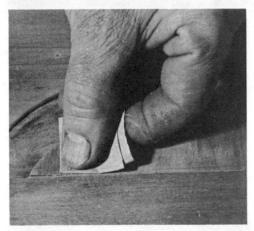

The first rule of sanding is "never across the grain." This is tough in situations where endgrain butts side grain. Handle the problem by placing the paper at the joint and pulling it away from the crack. When you're in the clear, start normal sanding methods.

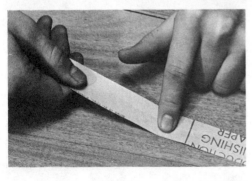

Very often you can blend out dents and mars in wood using sandpaper strips as shown here. Apply pressure with your finger over the blemish and pull the paper out with your other hand. Repeat, moving your finger slightly with each pull until blemish is smoothed out.

to a degree the grade of paper you end up with. Softwoods do not gain much smoothness from papers finer than 6/0. Hardwoods, the fine cabinet woods like mahogany, walnut, cherry, teak, oak, and some others, do become glassier under 8/0 or even 9/0 paper. Consider, also, the kind of finish as you turn to finer and finer paper. There is not much point in going above 4/0 or 6/0 for an enameled finish, since the pigments in the enamel

Shoe-shine technique on turnings and spindles is the fastest way to work. Follow this by working grainwise with finest grade of sandpaper.

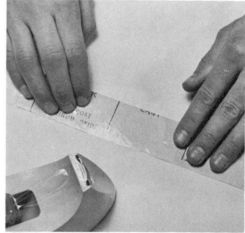

Give paper the strength it needs for shoeshine method by backing it with tape. This prevents tearing.

Abrasive cord is one of several ribbons, tapes, and strips you'll find handy for working in fine turnings and other odd shapes on furniture.

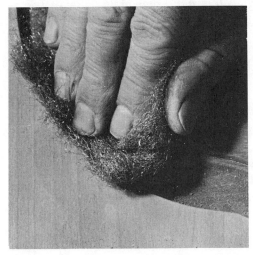

When cross-grain sanding is unavoidable use fine steel wool. It smooths the surface without forming sandpaper's minute scratches.

help you achieve the level surface you want. If the job is to be painted (in the purely functional sense) 3/0 is fine enough.

3. If the objective is a fine clear finish or a careful enamel finish, the next job is to dampen the wood with clean, warm water. The purpose of this is to allow the wood fibers at the surface to swell, as they always do in conditions of dampness. By deliberately forcing them to swell, before you put on the finish, you eliminate the chance that they may swell later, and make your carefully smoothed surface rough and uneven again. In addition, water raises the "whiskers" of the wood—the minute loose-end fibers —so you can sand them off. This dampening process is absolutely essential if you use a water stain; otherwise, the stain will raise the grain and give you a rough, uneven surface.

4. After the water has dried completely—overnight if necessary—sand lightly again with the same grade of paper you used for your finest sanding.

Important: If you are working on a much-sanded piece of old furniture with veneers, be gentle with the dampening, to avoid softening the veneer glues.

WHEN TO USE A SANDING SEALER. A sanding sealer enables you to sand wood smoother than would be possible any other way. When you turn to progressively finer grades of paper and the wood becomes smoother and smoother, it finally reaches a point where the abrasive raises a fine fuzz of fibers—and can do no more. Additional sanding merely raises more fibers. At this point, brush on a sanding sealer, if you need more smoothness. The sealer hardens and stiffens the tiny fibers and it "case hardens" the wood right at the surface. In effect, it turns the wood into a plastic-like material capable of taking a smoother finish.

You can buy sanding sealer at paint stores, or you can make your own by diluting shellac with three or four parts of alcohol. Brush it on smoothly, let it dry, and you'll sand the smoothest wood you ever saw.

However, *do not use sanding sealers indiscriminately*. The shellac or lacquer seals the wood somewhat, as you might expect, and this affects the penetration of stains and other finishing materials. Unless you need to control penetration, or the finish is to be enamel, or you don't care, run some tests on identical wood to determine the effect of the sealer on the materials which follow it.

VALUE OF A SANDING BLOCK. Nobody attempts to sand a surface smooth and *plane* without a sanding block. If you simply hold sandpaper in your fingers, it will sink deep into the softwood areas and ride high over the hardwood areas. When you finish, the result will be hills (hardwood) and valleys (softwood). If, on the other hand, you use a sanding block, it will give you the same result as a wood plane produces in cutting wood. The block—or the plane bed—rides the high (hardwood) areas, thus preventing the softer wood from forming valleys.

The most efficient size for a sanding block is about 4½″ by a little less than 2″. This size accommodates an eighth of a sheet of paper, cut down the center lengthwise, then each half into quarters. For bigger work, maintain the 4½″ dimension, but make the other about 3″. A block this size takes pieces cut six to the sheet. A block about 4″ by 5½″ takes quarter-sheets of sandpaper. (These figures are for garnet or aluminum-oxide paper; flint paper takes somewhat smaller sizes.)

You must not make the mistake of using a sanding block of plain wood. If you do, the hard friction between corners and high spots of the wood make the sandpaper heat up—make it clog and fill more rapidly. A thickness of felt (an old hat is perfect, or ordinary felt you can buy at dry-goods counters) on the block prevents this heating and gives you maximum sandpaper life plus excellent smoothness. The accompanying photographs demonstrate a quick way of making felted sanding blocks—a way so simple that you'll make several sizes, to meet the varying demands of different jobs. You can buy rubber, metal, and other sanding blocks. Try them if you wish, and see if you like their performance better than that of blocks you make for yourself. Generally, they are less useful because they are less flexible.

Not all sanding blocks are flat, because not all sanding is on flat surfaces. Make several sizes of round blocks, with which you can handle various diameters of concave curves. If you work on a project with a great deal of molding and other shaped work, you may want to tailor special sanding blocks to these specific curves.

In the accompanying photographs you'll see a method of interleafing two folded sheets of sandpaper, producing a pad which is four sheets thick. With this thickness of paper, there is less chance of the paper following the soft-and-hard curves of the wood. Use it for convex curves, where the curved surface forms the interfolded paper into an arch and gives you a very smooth job. Another method of sanding outside curves is with a sanding block made of a small piece of carpeting which bends to follow the

How to make a sanding block

Cut a scrap of ordinary lumber to accommodate an eighth of a sheet of sandpaper — about 4½″ x 1¾″. With gouge or rasp, cove the sides to provide a good grip.

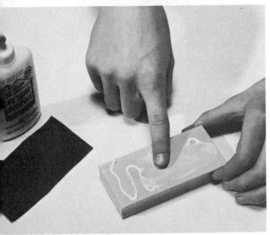

Rough-cut a piece of felt a little larger than the block. Smear white glue in a thin uniform layer over the block and smooth the felt in place. Avoid too much glue. If it soaks through the felt, the block is no good.

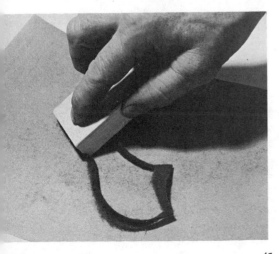

After the glue dries, remove the extra felt by drawing the edges of the block over a sheet of sandpaper. This bevels the felt and leaves a perfectly plane sanding surface.

surface. Some workers like a thin (half-inch or less) piece of foam rubber for a flexible sanding pad, although the rubber is not as good a guarantee against heating and gumming as the carpeting.

Not every surface, naturally, lends itself to sanding blocks. There may not be room for them. The grain direction may not permit their use. The situation may call for sanding with the paper handheld. Whenever it does, try to work with at least two thicknesses of paper, or with a thin piece of cardboard as a backup, to prevent cutting into soft areas.

Use power sanding equipment wherever possible, up to the final work with finest paper. Here the roller end of a belt sander quickly smooths the toothmarks of bandsawing. A basic sanding principle is in action here: always sand component parts of new construction whenever possible, since it is usually simpler than sanding them after assembly.

OTHER KINDS OF ABRASIVES. The abrasives used for sanding are not always in the form of sandpaper sheets, belts, and disks. You can buy long strips, in rolls, for use with special sanding blocks, or with blocks you make yourself in proper dimensions. There are various widths of abrasive ribbons, up to half an inch or so, as well as several sizes of abrasive cords. These cords and ribbons are usually coated with emery, and are intended more for industrial use than for the home craftsman. If the sort of work you do demands them, however, they are available from W. C. Mitchel, Middletown, Mass.

At hardware and paint stores you can buy a variety of abrasive products which are not in the form of sand and paper. Many of these are in the form of toothed metal sheets. These products last a long time, cut well, are slow to clog and easy to clean with brush or solvent when they do. You mustn't expect them to do finished work, however. Their function is in the early stages of sanding followed by fine sandpaper.

Follow power sanding with hand work if you're after real smoothness. Here a piece of closet pole covered with felt is used with a twisting, drawing motion to make endgrain glassy smooth.

STEEL WOOL. The action of steel wood is much like scraping. It produces minute, sharply cut shavings. As a result, the surface steel wool leaves is very smooth. It may not be level, however, since a wad of steel wool will cut into soft wood faster than into hard. You can control this almost entirely by using steel wool on a block. As it now comes, steel wool is almost in the form of a wide ribbon, in a roll. Unroll the ribbon a few inches and tear or cut it off. What you have is a "sheet" of steel wool, which you can handle like any other sheet. It will come apart fairly soon, but not until it has given you a very smooth surface.

Steel wool is best on hardwoods. It is at its very best on carvings, turnings, and other shapes which are difficult or impossible to get at with a paper abrasive. Buy it in several grades from medium to extra fine, and adopt a schedule similar to that you use with sandpaper: go over the surface with coarser material, gradually shifting to finer and finer.

If your fingers are soft, the tiny filaments of steel wool may prick the skin and break off. Wear rubber gloves to prevent this.

CHAPTER SIX

FIX-UP IS PART OF FINISHING

AN IMPORTANT part of wood finishing is the repair of natural blemishes and physical damage to the wood. Even on a new project, you can't always cut out or bury defects—and in refinishing work, you always discover dents, cracks, breaks, peeling veneer, and loose joinery. There is a definite order of operations and techniques which makes this sort of work simplest.

1. Reglue and clamp up all joints, drawer corners, stretchers, edge joinery, etc., that have come unstuck.

2. Before you stain or apply the finish, make all repairs that involve the use of wood. This includes such things as new rungs or stretchers, replacement of a corner broken off a drawer front or molding, etc. These wooden repairs should be in place when you do the actual sanding and finishing.

3. After you stain and apply the first coat of finish, make any repairs involving such nonwood materials as Plastic Wood, Duratite, Wood Patch, or the new vinyl putty. The reason you do these repairs at this time is to be sure *the color match is good.* Only after any stain and at least one application of the topcoat do you know what the final color will be. Then you can set out to match it with packaged colors of the patching material, or with colors you blend yourself.

4. After *all* finish is on and dry, you are ready to use the "wax pencil" type of filler, which can not be used under normal topcoats without interfering with drying.

An exception to this is the material known as "stick shellac," an old standby for repair of minor blemishes in such basic cabinet woods as mahogany, walnut, cherry, and some others. It is used on the bare wood, since it goes on hot and the heat would ruin any finish. When you use stick shellac, you must either anticipate the final color, or create a color sample with a scrap of identical wood.

REPAIRING LOOSE JOINERY. Whenever possible, completely disassemble any joint that is loose. Clean away all old glue, using a rasp, a scraper, or hot water on old hot glues. A good, modern glue surpasses any-

Dents, scratches, mars, blemishes that are not too deep often come out best if you blend them smoothly into the surrounding area. The advantages: no problems making patch match wood.

The first step is to use a sharp knife to smooth out the edges of the blemish. Within reason, make this gradual depression as large as possible, to keep the edge of the "dish" flat.

Finish the job with rasp and sandpaper. When a piece is quite old, this technique may bare some deep-down wood that is lighter than the surface. If so, a little gentle staining produces the perfect match.

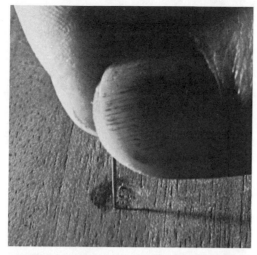

When no wood is missing — the damage is merely compression of the wood — cover the area with tiny pin-pricks, then soak it with hot water. This usually softens the fibers and most of the dent comes back to level. In stubborn cases, protect the area with aluminum foil, and apply a hot electric iron to the water-dampened area.

Classic treatment for common cabinet woods is the "shellac stick," in colors to match the woods. Stick, like sealing wax, melts in heat of alcohol lamp, and drips the blemish full of patch. You can use a match or cigar lighter — but their smoky heat usually turns the shellac off-color.

When the hole is overfull, compress the warm shellac with a gently heated knife — not too hot, or the shellac will bubble. Finally, shave it level with a sharp chisel, then sand.

thing they used even a short generation ago, but don't expect any glue to hold joints which show much of a gap when held tight. Very loose joints must be rebuilt by first gluing strips of sound wood to one or both members, then dressing them to a snug, gluable fit.

If joints are loose but you cannot get them completely apart, try injecting glue deep into the joint with a special device sold by craftsmen's supply houses, or with a long-snout oiler. A somewhat gimmicky alternative is to ask your doctor or dentist for an old hypodermic syringe to use as a glue injector.

It is very often possible to work glue into a joint by repeatedly tightening and loosening clamps, squeegeeing the material into the crack each time you open it.

Loose spindles, stretchers, and other socket-type joints will respond very well to such wood-swelling materials as Chair-Loc, unless the joints are so loose that the wood can't swell enough. In this case, you can often supply additional wood in the form of tiny wedges. Or, wrap the end of the spindle with twine, to supply required volume. Be sure to wash away any wood-swelling liquid which gets on the surface, since it may cause trouble with finishing materials later on.

REPLACING MISSING WOOD. Parts of a piece you refinish may be missing or broken beyond repair. Others may turn out to be wood of a different and lesser species. Many times a piece of furniture made long ago was put together of the most readily available wood species, and a chunk of cucumber shows up in a chest made mainly of cherry. Quite often, legs of nice old tables were of beech, while the rest of the piece was birch or maple. When you run into this situation, you may want to replace the odd-species member.

One of the first problems is finding the right wood. The standard cabinet woods are usually available in a good lumberyard. You can buy

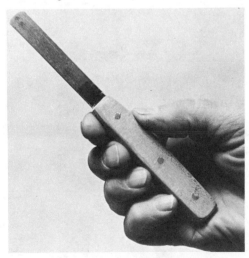

No tool is the crack-filling equal of an old hacksaw blade fixed up with a handle. Shortcut: make the handle out of tape. Closest to it is a palette knife from an art store.

Basic trick, using synthetic patching materials, is to buy a color close to what you want, then temper it, if necessary. This Lauan Wood Patch was picked for an antique pine chair seat.

As it goes on, the material looks dark. When you use it, be sure to compress it firmly into the depression being filled, to guarantee good adhesion.

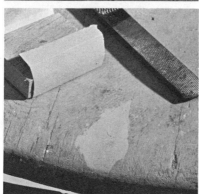

Smoothed and sanded, the patch turns *lighter* than the wood. This often happens with synthetic fillers — which turn dark again under the finishing materials.

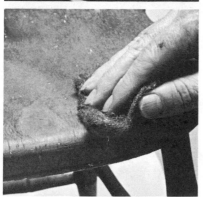

With a coat of penetrating resin for the desired color and to act as a sealer, followed by two coats of satin finish varnish, then steel wool, the patch is so close to the color of the old wood that you see it only if you look for it.

To prevent patches from loosening, give adhesion a boost by cleaning the crack carefully with a multiple-solvent intended for new wood. Swab out with cotton tufts, and let dry before you fill the hole.

Fill cracks with Plastic Wood in several layers, if the void is deep. Otherwise, the fill may shrink. Pile it a bit higher than the wood surface, to allow for some shrinkage.

Dress Plastic Wood fill with a sharp chisel, then finish it with sandpaper. Material used here is natural, to be spot-stained to match the surrounding area.

When the patch is hardened and smoothed, matching it to surrounding wood is simple, using oil stains, wiping stains, or water stains with some of the filling materials which are more absorbent.

virtually any species of wood from Constantine or E. H. Wild in New York or Craftsman's Wood Service in Chicago. However, some woods are easy to find around the house:

Hickory for chair spindles comes from an old ax handle.
Ash comes from an old rake handle.
Birch is used for most broom handles.
Maple dowels are common—as are birch.
Beech shows up often in cheap tool handles.

Since certain woods are difficult to find, and since it is often impossible to match up a newly sawn wood with one of the same species that is old, you'll find it a good idea to save odds and ends of uncommon wood.

When you make a new part, putting it in place may be difficult without more disassembly than you want to undertake. Very often you can solve this problem by making the new part in two pieces with a tapered splice. You insert the two halves, then glue the splice.

The tapered splice or scarf joint is essential in all small members, where the glue-line would be insufficient with a butt joint. Using a good epoxy glue, you can butt-glue fairly small parts, however, if the joint will not be subject to much strain.

It is simple to patch broken corners—even hopelessly burned or damaged parts in flat surfaces—if you pick a piece of wood which closely matches the grain pattern of the original. An important trick, however, is not to follow the edges of the piece in such repairs, but *to follow the grain direction*. If the inevitably visible joints run parallel to the grain insofar as possible, they will be less conspicuous. Blend and spot-stain as carefully as you can.

Wax sticks are available in virtually any wood color, to match blemished wood or plywood. You can blend two colors together, if necessary.

Tightening loose spindles is easy with Chair-Loc and similar materials, which are non-drying, and which swell the shrunken and compressed fibers of stretchers, etc. Use a Stanley Web Clamp to pull a piece snugly into shape, then apply the wood-swelling material. Or — go around it twice with a length of rope, and twist the rope.

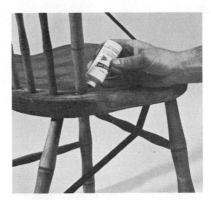

Make a steel dowel out of a finishing nail by pulling a spindle joint tight, then drilling through it, as shown here. Then drive in a nail. Combine this with Chair-Loc for a permanent repair job.

When stretchers and spindles have been neglected so long that the hole is enlarged, this trick tightens them. Cut tiny wedges and tap them in around the spindle. In many cases, a hole wears more on one side than the others; this can be corrected by wedging the worn side, thus truing up the piece.

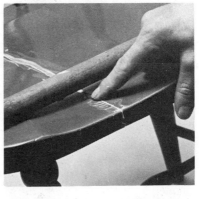

You can work glue deep into a loosened edge joint by rubbing it in with your finger, clamping it up, unclamping and rubbing the glue in again. Repeat this several times, and glue will be driven deep.

51

PATCHING AND REGLUING VENEER. Loose, lifting, or blistered veneer is simple to repair. Work some adhesive under it and either clamp the thin wood in place or hold it down with weights. Aluminum foil or plastic wrap do not stick to ordinary glues; they can be used to prevent adhesion to clamps and glue blocks. Use the smallest possible amount of glue, to avoid any unnecessary clean-up later which might damage the delicate veneer.

When the veneer is missing in spots, find the closest possible match in the same wood species—or some scraps of flitch from a throw-away piece of the same wood. Slice the wood as thin as you can, smooth the down side and cut it to a perfect fit. Glue it in place. Then, using a sharp plane and sandpaper, bring it down to the same thickness as the surrounding veneer.

One cause of difficulty in veneer patching is failure to remove the old glue. Old veneers are usually laminated with animal glue, which you can carefully wash away with warm water. To make this job easier, lay a bit of wet cotton over the old glue and soak it soft. Modern veneers, assembled with hot-plate adhesives, do not often fail. If they do, most available solvents won't touch the remaining adhesive, which must be carefully scraped away to the bare wood to insure adhesion when you reglue.

Be sure that the patch matches the grain direction of the original piece insofar as possible. You'll find that veneer repairs are less conspicuous when you cut the patch and the void into an irregular shape that more or less follows grain patterns. Rectangular or other recognizable shapes are the most conspicuous.

If you can find a good match in wood, the patch will finish out the same as the existing material. If there is much difference in color, however, do some gentle staining, to bring them together.

SYNTHETIC PATCHING MATERIALS. Plastic Wood, Rock Hard Putty, and similar materials are easy to work with if the finish is to be enamel—but they may be difficult under clear finishes when you must match a color. You can buy several brands labeled mahogany, oak, lauan, walnut, maple, etc., and sometimes the match is perfect. When it is not, you must blend and spot-stain, and you must not expect that a stain which you like *on wood* will be the same color on a synthetic patch. The only way you can get perfect results is with experimentation. Here are a few guide lines to help you along with color matching:

Plastic wood is hard and nonabsorbent. It will take oil stains to a degree, but reacts little to water stains. It's best to use a color that is close, then blend the shade if necessary. However, many good refinishers use only the "natural" Plastic Wood, and learn by experience how to tone it to the color they want. *Duratite* is quite similar to Plastic Wood, but a little more absorbent.

Wood patch is much more absorbent, and will take quite a bit of oil stain.

To clean away old glue in veneer repairs, lay a wad of wet cotton over the area until the glue softens. Then scrape it away. Clean wood is essential for smooth, tight repairs.

Put the carefully fitted veneer repair piece in place, cover it with a piece of aluminum foil, then back it up with a scrap of wood and clamp it. Foil prevents backup block from sticking. Best glue for veneer repairs is white glue; it's clean working and quick drying.

Simple scarf joint is best method of splicing new wood to a thin spindle. This provides enough glueline for a joint that won't break.

After glue is dry, use rasp, plane or other tools to shape new wood, complete spindle.

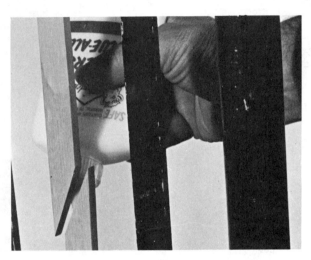

Angle joint here is used when a spindle or splat or slat must go between two rails and you do not want to take the whole piece apart. Shape the new piece and cut it a scratch long. Then cut it in two at a flat angle. Insert the ends in the holes and glue the joint.

Make new spindles or stretchers of hickory or ash, as required to match those in many old Windsor and similar chairs. For hickory, use an old ax handle. For ash, use an old rake handle. Drawshave or spokeshave is the best tool for shaping such a member.

Increase the length of legs when they are shortened by wear or breakage, with this technique. Cut the ends square. Prepare squares of matching wood and drill and counterbore them to go on the ends of the legs. Screw them on with a good, strong glue and plug the holes with dowels.

Then cut the square additions to shape with chisel, plane, rasp, or other tool and sand smooth.

If the joint shows — or if there is a problem in color match, you can minimize both by cutting a groove at the point where new wood meets old. If design suggests it, use two grooves. Four-in-Hand rasp is an ideal tool for this operation.

Rock hard putty, a cementitious water-mix material, takes both oil and water stains. It normally dries creamwhite, but some workers like to color it to wood tones with analine dyes as they mix it.

Vinyl patching materials are handy for their easy clean-up with water, the way you can smooth them with a dampened finger, and their smooth sandability. They are difficult to spot-stain, since there is little penetration, but you can blend them by using any of the wiping stains (such as Min-Wax) much as though it were a thin paint.

Although some of the synthetic fillers are labeled "non-shrinking," you'll have the best luck if you fill any very large cracks or voids with two or more layers, letting each dry before you apply the next. With all of them, make the patch a little high, then dress it to shape and level after it has hardened.

Another variety of patching-filling material is homemade, and in many ways very successful. Create a supply of sawdust of the same kind of wood you are patching. Mix up some *thin* Weldwood Plastic Resin Glue or some Elmer's thinned just a little with water. Stir the sawdust into the glue, to make a thick mash. Coat the hole to be patched with undiluted glue, then press the mash into it firmly. When the glue is dry, dress and smooth away the overage. This sort of patch offers a fairly stainable surface, once it is sanded smooth, and can usually be blended to a color almost precisely that of the surrounding area.

It is not always the best idea to patch. In many cases, a blemish can be blended so smoothly into the adjacent wood that, when it is finished, it is barely visible to the casual eye. An example is the fairly deep scratch in a table top. If you sand the scratch out, using *long* sanding strokes, the resulting depression is too gradual and too shallow to be seen.

In cases where the blemish does *not involve loss of wood,* but merely compression of the fibers and air cells, you can often bring the wood back up to level by soaking it with water. The action is more extreme if the water is hot. For greatest results, dampen the area, lay a piece of aluminum foil over it, and apply heat from an electric iron. In the most stubborn cases, use a needle or pin to puncture the surface of the wood in a closely-spaced pattern, all over the depressed area. Make these holes about as deep into the wood as the depression itself is deep. Then, water and steam. The holes will close up invisibly, and most of the blemish will disappear.

CHAPTER SEVEN

PENETRATING RESIN— THE EASIEST FINISH OF ALL

THE PENETRATING resin finishes, manufactured by dozens of paint companies and sold at every paint store, have gained a place among the most popular clear finishes. There is a remarkable list of reasons why:

- They are by far the easiest of all finishes to apply.
- They are among the most durable—so tough that they rank at the top in floor finishes.
- They have arrived with top designers in modern furniture, among whom the more natural a wood can look the better.

Penetrating resin finishes leave the wood looking entirely uncoated. This sample of teak has had the two-application system covered in this chapter, yet it shows all the textures of the natural wood. Despite this natural look, a penetrating resin finish is one of the most durable.

Ideal use for penetrating resin finishes is on small wooden objects, not only for furniture. It is best on woods whose texture is part of their beauty — walnut, mahogany, teak, rosewood, ebony, oak.

● They are equally proper on old furniture, even many antiques other than fine 18th-century mahogany.

● They give the wood a fire and color unlike that produced by any other finish. Wood hues are intensified—made more vivid—and there is no obscuring of the grain or wood texture.

● They are so foolproof and quick to use that they rank well at the top as finishes for large-area surfaces such as wood paneling.

● Penetrating resin finishes are like varnish—but instead of lying *on* the surface, they sink *in*. The resins harden between the fibers of the wood, in the air cells and other voids. When they are dry, the wood itself is measurably harder, and it is impervious to all ordinary household damage. In a sense they impregnate the wood with a plastic-like substance which turns the wood into plastic. But—there is no look of plastic.

WHERE TO USE PENETRATING FINISHES. Anyplace. They are resistant to alcohol, water, most household chemicals, heat, abrasion, scratching—almost any damage. Ease of application recommends them particularly for big furniture pieces. The simplicity of brush-on-wipe-off makes them ideal for work with much carvings or turning. They are the plain and simple choice for wood paneling, and have the greatest following of all for floors. Some penetrating resin finishes are intended *primarily* for floors, and these particularly rugged brands are superior for furniture, too. Penetrating resin finishes are not meant for outdoor use.

HOW TO APPLY A PENETRATING FINISH. Flow penetrating resin on the wood thickly. If possible rotate the work so that you can apply it on horizontal surfaces. Don't worry about fancy brush work. Many finishers don't even use a brush. They mop the material on with rags—or pour it on, then spread it. You can even use your hands, if you want to, rubbing the finish in as you spread it. Some experts use a pad of very fine steel wool to spread the puddles of finish and work it into the wood; this is one of the best techniques of all.

Keep the surface wet for about half an hour. Read the label, because some manufacturers ask for a full hour of penetration. If you see dull spots appear, indicating that all the material has soaked in, apply more to keep the surface wet.

After the wood has soaked up all the liquid it will take, use plenty of rags to wipe *all the surface liquid off*. Wipe it clean. Do not leave a trace of the finish on the surface, for if you do, it will dry to an unpleasant sort of sheen. Should this happen and you catch it the next day, you can usually cut the dried material by brushing on more finish, then wiping it in a few minutes.

The penetrating finish continues to soak in and permeate the pores and spaces in the wood after you wipe it. For that reason, a second application is usually a good idea, especially if the wood is porous. *Do not wait until the*

Basic steps to a good in-the-wood finish

Carefully, work from 3/0 up through 6/0 or finer sandpaper, using a felt-covered block. If it's a refinishing job, be careful not to cut through old-wood color, which darkens beautifully under penetrating finishes.

Dust thoroughly. Use a vacuum or a good dusting brush. You needn't worry about dustmarks on a penetrating finish, but you must get the wood pores clean, or you'll lose the good texture you want.

A good way to apply penetrating resin finishes on horizontal surfaces is to pour it on. Whenever you can, turn the piece so that the surface you're working on is flat. This lets the finish stand on the wood and penetrate as deeply as possible. When you use a brush, lay it on as thick as possible without runoff.

Swab the finishing material around on the surface with a wad of fine steel wool, using light pressure. This helps to work the finish into the wood. Keep the surface wet for fifteen minutes or longer, adding more finish if it is required.

Finally, wipe absolutely all the finish off the surface. You want it in the wood — not on it. After about half an hour, you may have to wipe again, since sometimes tiny spots of finish come to the surface and leave shiny spots unless wiped away.

Antique Cherry

Pine

Rosewood

Korina

Walnut

Pecan

Cherry

Fir

Penetrating resin finishes leave wood with a natural finish, but darken it slightly. These color samples show you the results on 16 different kinds of wood.

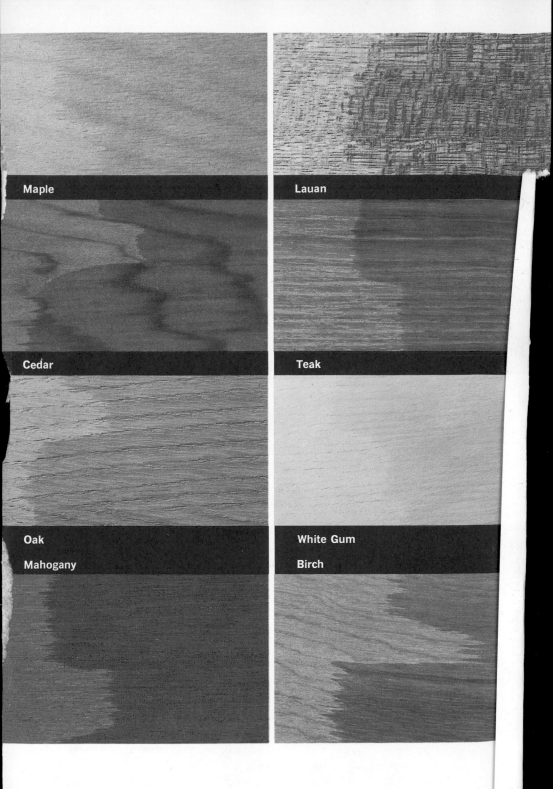

Maple

Lauan

Cedar

Teak

Oak

White Gum

Mahogany

Birch

first coat has dried and hardened completely. Three hours or so is long enough to wait. Brush or swab on the second coat. Keep it wet for half an hour or longer. Wipe it off. Again—*wipe it all off.*

When this well-wiped application is dry the job is done. You can wax the wood if you want to—but this tends to clog the pores and steal some of the natural, textural beauty of the finish.

THE COLORS OF PENETRATING FINISHES. The natural look of the penetrating finishes is at its best on open-pored woods such as oak, walnut, mahogany, pecan, chestnut, etc. It leaves their texture unfilled, and makes them look like wood. Maple, birch, cherry, and other close-grained woods take the finish, of course, but are not so spectacular as the coarser-textured species.

As shown in the color photos, the penetrating resin sealers tend to darken the wood somewhat in the process of intensifying the colors and emphasizing the grain pattern. The degree of darkening increases with natural darkness of the wood. For example, light-colored maple is darkened only slightly. Darker-colored walnut reaches its customary dark brown color. Rosewood's browns go almost black. This degree of darkening is considered ideal for each species. Stain is not usually required. If you want a penetrating finish that is lighter in color, use a wood bleach to lighten the natural color a shade or two. Of course, no filler is used with a penetrating resin finish.

Although the color given wood by a penetrating resin finish alone is usually appreciated by most people in the darker woods, you may want to stain some of the lighter species. If you use a pigmented wiping stain (see Chapter 8) then follow it with a penetrating finish. The finish will not change the color of the stained wood, other than to intensify it slightly. The reason for this is that most pigmented wiping stains are mixed in a vehicle that is much like a penetrating finish.

If you use a water or a non-grain-raising stain, however, you must expect a considerable intensification of the color, a shift in hue toward the reds, and some darkening. For that reason, it is best to make test patches, as discussed in Chapter 8.

HOW TO BUY PENETRATING FINISHES. There are two kinds, basically. One is formulated on the phenolic resins, the other on alkyds. The phenolics tend to penetrate deeper. Look for the word *penetrating* and the word *resin.* Or—look for instructions which tell you to brush it on and wipe it off. Every major paint manufacturer has a penetrating resin finish. Some examples are Watco, Deep Finish Firzite, Clear Rez, Clear Minwax, Dupont's Penetrating Wood Finish, and many more. There are materials that are similar in application based on oils, not resins. In situations where protection is not important, they can be used, although they offer no advantages in service or application ease.

CHAPTER EIGHT

WHAT GIVES WOOD THE RIGHT COLOR

THE CLASSIC, accepted, popular finished color of most woods is not the natural hue of the wood. Most of the time it is the color the wood *turns when it ages* and oxidizes and reacts to light. When you refinish an *old* piece of furniture it already has that color naturally, unless you are careless and sand it away. When you refinish a *recent* piece, it may have the color in the form of stain applied by the manufacturer, unless in your stripping and clean-up you removed it. When you finish a project made of new wood, you must stain it if you want the classic, accepted, popular finished color. And most people do. That is why most finished wood you see has had its color controlled to some degree through the use of stains.

WHEN TO USE STAIN. Many woods, including mahogany, walnut, cherry, teak, and other dark-hued species look wonderful in the color they take on from oils and resins in varnish, shellac, and lacquer. To many an eye, this color is especially beautiful when the finish is a penetrating resin. Stain is not usually needed for *aged* pine, maple, birch, and some others. Very few wood colors are as beautiful as the color of pine when it turns brown with age. The same is true of oak, maple, and birch. Another superb pine color is that which results from out-and-out weathering. The wood turns a dirty gray—then the finish turns those grays to deep red-browns with highlights of red-gold.

On the other hand, some woods, including new pine, maple, and birch, plus gumwoods, beech, poplar, and similar light-hued species need stain to give them any character at all. Philippine mahogany, lauan, larch, and similar middle-toned species can be finished with or without stain. Usually the proper stain gives these materials a little more life than they normally have.

As the color photographs of stained wood samples show, stains serve four different functions as they alter the color of wood:

1. They may be used to give wood the traditional color expected of it. Mahogany is a bit more pink in the raw than its popular hue. Walnut in the raw is more blue—less brown—than we like to see it. There are accepted colors for maple, birch, cherry, oak, pine. Stains impart to run-of-mill woods the colors they are "supposed" to be.

2. Stain may be used to give wood more character than it has naturally. Some perfectly good cabinet woods are simply prettier with a stain of some sort. A good example is the lauan shown in the color photos. Lauan is okay natural. With a little fruitwood stain it becomes more stunning than okay. Oak reacts the same way.

3. Stains may be used to give woods a better color sooner. Pine is the best example. Minwax Early American or Ipswich Pine or Puritan Pine— or similar colors by other manufacturers—put a hundred years on a pine board in about half an hour.

4. Stains are often used to make one wood seem like another. Cucumber or gumwood, relatively uninteresting in natural color, can be made the same general color as mahogany, walnut, cherry, or other cabinet wood. Never does this subterfuge come off entirely, but you've seen a lot of furniture with mahogany tops and fronts, but with aprons and side panels of another wood. If you didn't look closely, you didn't know the difference.

HOW TO CHOOSE THE RIGHT COLOR. There is, of course, only one right color for a given wood in a given project, and that is the color you yourself like best. The more *good* colors you see, the better your own judgement will become, so observe wood hues carefully in furniture stores, at antique shows, in the magazines, and other places where the best is on display. Carry the memory of these colors with you when you go in to buy stain.

Not many years ago it was a fairly common practice for woodcraftsmen to mix their own stains from pigments. It is still an interesting process, and it may be the only way you can find the exact color you want. But—there are dozens of manufacturers of stains of one kind or another, and since their colors are not identical, you have a choice of ready-mixed hues that is almost limitless.

Some of the better stores will work with you on a stain color until you find what you want, using the pigments of the standard custom color system. There are wood-finishing services, operating by mail or direct contact, which specialize in color matching. You send them a chip of the color you want and tell them what wood you want to put it on. They will work out a complete step-by-step finishing schedule and send you a sample of the wood with each step shown. One such company is Gaston Finishes, P.O. Box 1246, Bloomington, Indiana.

The next chapter covers mixing and modifying stain colors as may be necessary to give you the exact colors you want.

CHAPTER NINE

HOW TO USE WOOD STAINS

STAINING can be the most meticulous and fussy work in the finishing process—or it can be a merrily slapdash operation you could do with your eyes closed. The difference lies in the kind of stain, the nature of the project, the variety of wood, and the condition of the surface. In no circumstances is the work as difficult as it once was, because today's stains and techniques eliminate blotchiness, overtoning, lapmarking, and other problems. This chapter will tell you the few simple tricks that make staining a sure process, not a guessing game.

KINDS OF STAINS. Ninety percent of your staining needs can be taken care of by three materials: penetrating wiping stains, water stains, and non-grain-raising (NGR) stains. These and others are covered individually below.

Pigmented wiping stains. The prototype material in this classification is Minwax. Essentially identical stains are made by virtually every paint manufacturer, with names like "oil stain," "wood finish," "wood stain," etc. These stains are a penetrating oil-resin vehicle in which *pigments* have been stirred. The pigments are not in solution, but in suspension, and the material must be stirred rather constantly to maintain a uniform color. There may also be some color in the form of *dyes* dissolved in the liquid, but the major job of coloration is done by the solid pigments.

This family of stains includes some in a water emulsion (Deft is an example) and another that comes in a tube—toothpaste thick—and is one of the easiest of all stains to use (Even-Even.)

The wiping stains tend to leave the most pigment in pores, in softwood areas, in blemishes. This means that they greatly accentuate the grain pattern, as well as any places which are not properly and smoothly sanded. Woods that are soft and porous, such as pecan, lauan, fir, etc., accept a great deal of stain, darkening greatly. Harder woods such as maple, birch, beech, etc., take less stain. Walnut, mahogany, oak, and similar hardwoods with open pores accumulate a lot of the pigment in their pores.

Pigmented wiping stains are ABC simple. Brush them on, let them penetrate for a few minutes, then wipe off. Penetration time controls darkness of stain. Be sure to wipe clean — or the pigment tends to obscure the wood.

The final color of a finish comes from the stain plus the topcoating. Here a varnish is being brushed over test swatches of water stain on walnut. Each shade is a weaker dilution with water from straight to one-to-one, two-to-one, four-to-one, etc. If your project involves turnings, make tests on a prototype turning.

Although pigmented wiping stains come in all the regular woodtone colors (plus more) and are regularly used on any and all woods, they perform their best magic on pine and other softwoods. Owing to the way cracks, dents, and scratches catch the pigment, wiping stains are excellent for any "distressed" finish, as well as for furniture and paneling in the "primitive" style.

Application of pigmented wiping stains—covered in detail later on—is very simple.

Water stains. These are true stains, in that they *dye the fibers of the wood* the way cloth is dyed. The colors are brilliant, stable even in sunlight. Penetration tends to be uniform, even in woods that have widely varying porosity, such as fir plywood.

Water stains are very inexpensive. For something like 30 cents each you buy packets of dye powders. These usually have wood-hue names and are blends of aniline dyes produced by a chemical dye house. Also, there are pure colors—black, red, green, etc. You dissolve a packet of dye powder in a quart of water close to boiling. This forms a stock solution which you may use straight—dilute with water for lighter shades—or modify with other water stains. Not all paint and hardware stores stock these dry powders, but they can order them for you. Or—you can order them yourself from such woodworker's supply houses as Albert Constantine.

There is much variation in stain colors from one line to another. A packet labeled "walnut" may be three different colors from three different manufacturers. For this reason you may have to experiment a little to find the colors that please you best. Constantine's stains have been found to be very close to the most popular conception of what the wood colors should be. Some dye powders are labeled "water & alcohol soluble," and as the name suggests will dissolve in either water or alcohol or both. These powders do not produce the same color in alcohol (which produces cold, somewhat greenish tones) as in water (which produces warm, reddish tones). One of the most interesting and subtle blendings of stain colors comes from varying proportions of water and alcohol with aniline powders which dissolve in either or both.

Water stain comes in the form of aniline dye powders. An ounce mixes into a quart of almost-boiling water. Some powders will dissolve in water or alcohol, or a mixture, giving you interesting differences in color as well as in drying time.

Water stain powders usually carry woodhue names. With four or five of them, you can blend almost any shade you want. Add black to darken a shade and to kill too much red. Use careful measuring techniques, record proportions so you can repeat.

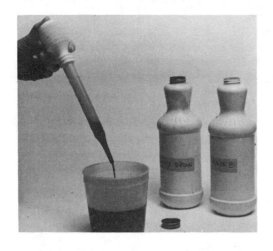

To mix small quantities of water stain — or to maintain careful control over quantities—try a plastic basting syringe at housewares counters for about 50 cents. Some of them are calibrated in fractions of ounces.

Non-grain-raising stains. Stains which do not use water—or use only a relatively small amount of water—do not cause the grain to raise, since the liquids they employ do not soften the fibers. NGR stains (sometimes called NFR, for "non-fiber-raising") based on chemical by-products of the petroleum industry have about the same qualities as water stains, but dry for topcoating in a few hours. The colors are a bit more intense than those of water stains when they dry—but they are close to identical in brilliance under topcoatings. The NGR stains are considered, largely, to be industrial products and for that reason they are hard to buy at retail. You can order them from Constantine or from Craftsman. They come in quart bottles (or bigger) and it is a good idea to order thinner at the same time; you'll rarely use an NGR straight. These stains can be thinned with wood alcohol, if necessary, however.

Stains, varnish stains, and stain lacquer as shown here are easy to mix. Work with the fewest possible pigments. Burnt sienna, for instance, produces good color variations with the addition of varying amounts of black.

You can make your own NGR stains, if you wish, using alcohol soluble powders and wood (denatured) alcohol. This stain will dry a bit faster than commercially prepared NGR, but you can slow it down by adding a *small* quantity of water. Although this would seem to destroy the non-grain-raising properties, the small amount of water does not raise the grain enough to become a factor in the finishing.

NGR stains are not recommended for pine, fir, spruce, and similar woods with wide variation in density between spring and summer growth, since the grain pattern may go wild. They are, however, high favorites on the standard cabinet woods and are pretty much the universal stain used in furniture factories.

To mix a good-working pigment stain, use a penetrating resin sealer or a resin-oil blend as the liquid, and stir in fine-ground pigments. Basic colors are burnt and raw umber and sienna plus red, blue, yellow, orange, and green. Black is a key color, used to darken too-bright hues.

Oil stains. Although some pigmented wiping stains are labeled "oil stain," they are not true stains in that they contain solid pigments. True oil stains are composed of an oil-base vehicle in which oil-soluble dyes are dissolved. They penetrate the wood and permeate the fibers with color. These materials are not very common, appearing nowadays mainly in products which closely resemble *colored* penetrating resin sealers.

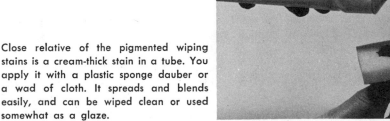

Close relative of the pigmented wiping stains is a cream-thick stain in a tube. You apply it with a plastic sponge dauber or a wad of cloth. It spreads and blends easily, and can be wiped clean or used somewhat as a glaze.

Varnish stains. Some surfaces never invite a close look—such as insides of cabinets, backs of chests, bed rails, and the like. These may not merit the bother of the same finish you'd put on conspicuous spots. A coat of varnish stain in a closely matching color is a worksaver. You brush on varnish stain just as you would regular varnish. Since there are oil-soluble dyes in the material, it leaves a film on the wood which imparts a color while it obscures the wood grain to a great extent. For this reason, varnish stain works when you are using an inferior wood and want it to take on the same general hue as a better species. It saves the need for stain plus varnish or other topcoat.

Padding stains. Least known among home craftsmen is a type of stain which you apply with a wad of cheese cloth or other rag. These stains are never put into liquid form. You moisten the rag with thinned shellac or a dilute lacquer known as "padding lacquer." You touch the rag to a little colored powder and rub it over the wood until the color is spread about, usually in a pattern of shading somewhat like natural variations in wood tone.

Padding stains are *intended for use over old finishes,* and one of their major uses is for touch-up. Antique dealers often use them with burnt umber to put down a surface layer of pseudo-antiquity which is convincing

Padding stain is a special powder you apply with a cheesecloth pad dipped in very thin shellac or special padding lacquer. It is intended for touch-up only — never on bare wood — and is widely used to liven the color of old finishes when complete removal and refinishing is not desirable.

enough to fool a fair judge of old furniture. Another trick is to give an off-species of wood (a cucumber panel in a cherry chest, for instance) a color that more closely matches the major wood of a piece of furniture.

Since the liquids used for padding are not of great durability or resistance to moisture, alcohol, etc., they should have a topcoating of varnish or other protective material.

Walnut sapwood stain. This is a specialist material used to bring whitish sapwood of walnut down to about heartwood color, after which the entire area is stained. Craftsman Wood Supply sells it. Also—you can experiment a bit and formulate your own in any of the standard types of stain.

Chemical stains. Certain standard chemicals will cause a discoloration of wood, and when the color is a desirable shade, the result is good looking and completely fadeproof. Most useful of these—and you may want to experiment with it—is potassium permanganate, which your druggist may not sell you since it is poison, but which can be bought at a good photo shop, since it is used in photography. It turns wood a natural, oxidized-aged brown color. Another useful chemical stain is copper sulphate, which turns wood black. The easiest way to use chemical stains is to mix a saturated solution, then make tests with weak dilutions. The trick is to apply the permanganate in a complete flooding. Avoid trickles. Shoot for uniformity of application and of time lapse, since the material works as long as it remains on the wood—until it reaches exhaustion. The problems of uniformity eventually lead most wood finishers to give up chemical stains.

Bleaches. Although they are the precise opposite of stains, in that they remove color from the wood instead of adding it, wood lighteners belong in any discussion of wood color control. They are covered completely in Chapter 10. There are many chemicals (oxalic acid, Clorox, hydrogen peroxide, etc.) that will bleach wood; you'll get the best results most dependably from such two-solution products as Blanchit or Albino bleach. Two-solution bleaches are foolproof. Following instructions on the label, you apply one solution which sets up the wood, chemically, for the second solution to take away the color. With one treatment, a dark wood becomes pale. With two, it becomes pure white. Follow with a clear lacquer, to prevent yellowing.

STEPS TO A GOOD STAINING JOB. Before you open a container of stain, remind yourself that you can't do any more sanding or other preparational work after the stain goes on—except for certain patching which must match the final color. For that reason, check these points:

1. Be sure the wood is smooth or be ready to accept the consequences of roughness. Stain accentuates roughness—even the minor scratches of cross-grain sanding. There are times when you don't mind—perhaps even prefer—some roughness to be picked up by the stain. But if you don't want blemishes to show, get rid of them.

2. Be sure the wood is clean. Good uniform staining is impossible on a dirty surface, since the dirt not only may show, but it may retard penetration of the stain. The best way to be sure of a clean surface is to wipe the surface carefully with a product such as Cleanwoode, or lacquer thinner.

3. Be uniform. Don't sand one area with 3/0 paper and another with 6/0. The relatively rougher 3/0 area will stain darker in many cases. Don't make the mistake of expecting a surface that has gone through a jointer or planer to be ready for stain. These tools have a burnishing, polishing effect on the wood which seals off some of its ability to absorb. These surfaces must be sanded.

4. Seal end-grain which is sure to overabsorb. Woods such as fir plywood should be sealed so that their soft and hard areas will accept stain more uniformly. On end-grain use a washcoat of shellac—about 4 parts of alcohol to 1 part of 4-pound shellac. Highly porous end-grain may take a richer mixture. Brush this washcoat on carefully, if you want uniform end-

grain staining. You will find that end-grain sanded perfectly smooth with 6/0 or finer paper does not overstain, particularly with water and NGR stains. The technique of sealing soft-and-hard areas is covered in the discussion of fir plywood, in Chapter 17.

HOW TO APPLY STAIN PROPERLY. Before you stain any wood, you must establish the color you want. Most of the time, this will be a standard woodhue stain, mixed by the manufacturer. If you need to change a standard color, you'll find the techniques covered in the accompanying photographs.

To make a stain darker, add black—making sure that it is a compatible color medium. Black water stain, black NGR stain, and black (ebony)

Although a cardinal rule for finishing is to stain over clean wood only, now and then you slip. When some sort of dirt holds a stain out, try rubbing it in with a small piece of steel wool. The slight abrasion helps the stain penetrate.

pigmented wiping stains are available. Use them to darken a stain. In the case of pigmented stains, use black pigment you buy in tubes.

To make a stain lighter, you have a choice of four methods. First, you can dilute the stain. Second, you can wipe it quickly, before penetration is great. Third, you can washcoat the wood with thin shellac to reduce penetration. Fourth, you can pretreat the wood with the solvent used in the stain, following quickly with the pure stain. This method is a variation of dilution, since the solvent already in the wood tends to weaken the stain. It is not as easy to control as the first three methods, and therefore most finishers hesitate to use it except for projects so small that they can be handled in seconds with quantities so small that diluting them would be wasteful and difficult.

NGR and water stains do not always require wiping. For that reason, the most effective means of controlling their depth is through dilution.

Penetrating wiping stains may be diluted with a penetrating sealer or a product such as Val-Oil. This reduces the amount of pigment deposited on the wood. Then, you have a further control by wiping sooner.

The final color of a stained finish is the result of the stain *plus* the topcoat. Usually the color left by the stain alone is a little flat, and a little light. When you put shellac or varnish or a penetrating sealer over it, the color intensifies and darkens. At the same time, it turns slightly warmer—that is more reddish. When you topcoat with lacquer, the intensification is the same but the darkening is slightly less. Meanwhile, there is little effect on the warmth of the color.

Since these factors must be taken into consideration, you must experiment with these stains if the final color is critical and without tolerances. Use a sample of the identical wood, if possible—or one that is very close. Carry each experimental swatch through to topcoating.

Actual stain application can be by any common method commonly used to put a finish of any kind on wood. You might as well dip a tiny project; you might as well roll a paneled wall or a floor. Spray is commonly used for water and NGR stains. Here are the basic techniques, by stain types:

Wiping stains. Although you may wipe all stains to a degree, wiping is the *key step* in the use of pigmented wiping stains. Put it on any way at all, as long as you get complete coverage. Brushing is easiest and usually neatest. Allow the material to penetrate. It is standard to wait until it starts to dull over. However, this produces the darkest coloration; wipe sooner if your color requirements demand it. *Wipe clean.* It will be impossible to remove all the pigment from the surface of the wood, but you should try. What remains in the pores and fine, almost invisible scratches provides the color.

If, after you have wiped all you can, the color is still too dark, dampen a cloth with paint thinner and *carefully* wipe more. This same technique is used in "blending"—that is, lightening certain areas which may go darker than the rest of the wood.

Blending is more often used, however, to *darken* portions of the wood which are too light—such as sapwood. When you work with pigmented wiping stains, darken these places with a second application of stain, working carefully to blend out the edges so the patch doesn't show. You may have to dilute the stain to prevent overdarkening, and will most likely find it necessary to wipe selectively, leaving a bit more pigment here and there —or a bit less—to equalize the color and produce a natural look. (Blending is done first when you use water and NGR stains.)

Pigmented wiping stains provide a degree of protection in themselves. Their vehicle is often a penetrating sealer formula. Therefore, they may need no more than a single coat of varnish or other topcoating for a complete finish. On the other hand, if you want a fine finish, rubbed and brought to a good surface, it is best to forget that there are resins in the stain. Treat it as an oil stain. Give it twelve to twenty-four hours to dry before you proceed with the finish. It is not necessary, however, to seal it with shellac as many finishers seal oil stains, since there is no tendency for the pigmented wiping stains to bleed.

Water stains. It is unfortunate that more people do not take advantage of water stains—their economy, permanence, good colors, and flexibility. They have only one drawback—which should not deter anyone in their use:

water stains tend to raise the grain of the wood. Softened by the water, tiny fibers stand on end. Minute places at which the wood may have been slightly compressed by the pressures of abrasives tend to rise again, producing slight roughness. If this roughening is severe enough, you can not sand it smooth *without sanding away some or all of the stained wood.* However, very slight roughening is smoothable, if you use 6/0 or finer paper and just whisk over the surface.

The proper defense against grain raising, however, is to dampen the wood deliberately after it is entirely smooth. Then do the once-over-lightly sanding with 6/0 before you stain. The grain will not raise again. This trick is even more effective if you pre-dampen with warm water. (See Chapter 5, on the use of sandpaper).

Water stains produce more uniformity and less distortion of the wood-grain pattern than any other stains, and for that reason are very easy to use. You simply flood the surface with stain. Let it dry unless it seems to puddle, in which case you merely rag off the surplus. Since the water tends to occupy the wood pores and interfibrous spaces in about the same degree as the tree's sap did, the relative coloration of lighter and darker woods is about directly proportional to the natural lightness and darkness. This gives wood an honest look, much appreciated by fine finishers.

Since there is no sealing of the wood with water stains, you can put the stain on in successive coats, if you wish. This is an excellent procedure. Dilute the stain quite a bit more than required. Apply a second coat. When it is dry, put on another. Then, if necessary apply a fourth. Each application produces the same degree of darkening. You can go as slowly as you like, with no danger of overstaining.

Some of the following tips will be helpful:

1. To prevent water stains from bridging wood pores in a material like oak or chestnut, add a few drops of Kodak Foto-Flo or other surface-tension reducer which "makes water wetter." Otherwise, take pains to work the stain into the pores, or they may come out light. This is of course no problem if you are planning to fill the pores anyway.

2. Turn the piece as you work so you are always applying stain to a horizontal surface insofar as possible.

3. Work with a good-sized brush on big areas so you can keep the edge wet constantly. Otherwise, puddling may dry slightly and show rings.

4. Try to work on individual areas—a top, side, drawer fronts, etc., and be careful not to slop over on other surfaces until you are ready to handle them. This avoids trickles that may show dark later on.

5. When you stain large vertical areas, work from the bottom up, so dribbles run into prestained areas and can be wiped before they overstain.

6. When you stain carvings and other raised places, gently wipe the high spots so they are lighter. This gives emphasis to the carvings and helps maintain the proper aged look in period and antique furniture.

7. If water stain is too dark, you can raise it in tone somewhat by sponging with warm water. Use plenty of clean rags.

The proper method of blending lighter areas to match darker wood when you use water stain is to dilute some of the stain to the proper intensity so that it *makes the light wood the same color as the darker*. Then, when you stain the entire piece, the lighter wood disappears entirely. This trick is easy to do with a little practice, and is the method used by professional finishers. It is even easier if you spray the stain.

NGR Stains. These materials are the closest of all stains in contemporary use to the old spirit stains. They strike deep and hard into the wood, dry quite fast, and are less tolerant of clumsiness than water stains. However, they do not raise the grain—which eliminates one problem. They dry for recoating in about four hours—which eliminates another.

Handling NGR stains is basically the same as water stains, except for their faster drying. You must be a little more in control of the situation to avoid lapmarks, runs, and other unevenness.

WHICH STAIN FOR WHICH WOOD? Any stain can be used on any wood, of course. However, there are certain combinations which work out best.

Pine, fir, spruce, larch, and other conifers are excellent with pentrating wiping stains. Their porousness traps just the right amount of pigment. The oils and moderate dye content darken and tone the wood just the right amount.

Maple, birch, beech, and similar hard, close-grained woods take excellent color from NGR stains, especially if you want to tone them fairly dark. Water stains are good on these species, too, but their slow absorption means slower drying time and easier handling. Some of the best maple colors are in the NGR stains. Oak is another wood which takes good tones from NGR.

Mahogany, walnut, cherry, and other woods which normally require only a slight toning are at their best with water stains. NGR works just as well, but in the hands of the average worker, water is easier to dilute and control, and may need just a touch of stain on walnut or mahogany to give it a perfect color.

CHAPTER TEN

HOW TO LIGHTEN WOOD

STAINING, as covered in the preceding chapter, alters the color of wood by *adding* color. There are times when you want to *remove* color from the wood, making it lighter. Throughout the early history of cabinet making, this was not possible. Sheraton, Hepplewhite, Chippendale, and the Adam brothers might have yearned for a method of making wood lighter than it is, but their technology didn't provide the means, although they did use naturally light satinwood in those days. The chemicals and the techniques of wood lightening did not become popular to wood finishers until the 1930s, when "Swedish modern" came into vogue. In that era of furniture styling, it was common to use a strongly white-pigmented filler. When this pigmentation was applied to various pale woods such as prima vera, the result was a sort of escape from the traditional colors of mahogany and walnut and cherry.

The period during which much furniture was finished "blonde" was short-lived, and wood returned to its traditional colors. Nothing like the blonde craze has ever returned in furniture, yet there are occasions when you want to make wood lighter than it is. Sometimes woods which for reasons of texture, structural strength, availability, or fashion are considered best suited to cabinet making are darker than you prefer. It is simple to make them lighter, and there are several ways of doing it.

WOOD BLEACHES. A bleach turns the wood lighter by removing some or all of its pigmentation. Walnut turns creamy in color. Mahogany turns a very pale rose. Wood lighter in native color may turn quite white. The general tone of the wood is retained. Only the color intensity is reduced.

The best and most thorough-working bleaches are those you buy at paint stores, usually in two solutions although one-solution bleaches are available. The standard procedure is to brush on the first solution, which changes the pigments chemically so that the second solution removes them. Some bleaches can be mixed into a single solution for somewhat lesser effect.

Comparative lightening of wood using bleaches: Two-solution commercial bleach on left lightens walnut plywood this much in a single application. The darker portion of the sample is untouched. Household bleach in two applications at right produced the lighter shade. This sample is also of walnut plywood.

Before you attempt to bleach wood by any means, be sure that it is clean and free from oils, grease, old finishes, and any other dirt which might hold the bleach out. Remember, the bleach is an aqueous solution; anything that repels water will reduce its effectiveness and give an uneven result. You must keep in mind also the grain-raising effect of water. The fibers of the wood are softened more by bleaches than they would be by plain water, and they stand up in tough whiskers. If you dampen the wood before you bleach, then sand off raised grain, you minimize this effect. However, since the bleach has a stronger grain-raising characteristic, you still get some roughening. Usually the depth to which the bleach is effective permits enough sanding to restore smoothness with or without predampening.

Although it is the second solution in two-solution bleaching which removes the color, careful application of the first solution is the most important. Flow it on freely and be sure the surface is well saturated. Try to wet the surface uniformly. Sometimes you may find that the wood resists wetting, most often when you are working on an old piece from which you've removed the finish. When this happens, use very fine steel wool to help the bleach work its way into the wood.

The first solution usually takes about fifteen minutes to work, but it does no harm to leave it longer. Even if it dries, you still get your bleaching. Brush the second solution over it, again making sure that the wetting is complete. In about four hours—depending on drying conditions—the liquid evaporates and the wood is bleached. In actual practice, it is better to let the work dry overnight, as would be the case with any dampened wood. If the weather is damp and muggy, put off the bleaching until another day.

Lacquer lightener takes advantage of the non-oil character of lacquer, which doesn't darken wood greatly. After the lacquer is applied, to seal the color, you can use any topcoating material without any change in the color of the wood. This enables you to use the extra moisture and abrasion resistance of some oil-resin finishes without their deepening the wood hues.

If you want the wood whiter than this application of two-solution bleach makes it, repeat the entire operation. Do not sand between the two treatments.

There are small variations in application techniques between brands, so be sure to read labels carefully. This applies particularly to mixing the two solutions for single application. Usually you use two parts of the first solution and five parts of the second.

Precautions: The solutions are strong. Wear rubber gloves to prevent the possibility of skin irritation.

Use a nylon brush. Rinse it after you put on the first solution.

Pour the solutions into glass containers for use, and do not pour any unused solution back into the bottles. Metal containers are likely to cause discoloration of the material.

OTHER WOOD BLEACHES. You can bleach wood with several different chemicals often found around the house or in the shop. Ordinary Clorox lightens wood quite a bit. If you use several applications, the effect is cumulative and you can take out a lot of color. Don't sand until the final application is dry.

Oxalic acid, mixed in a strong solution, takes pigments out of wood and also bleaches some stains that may have stayed in the wood after stripping. The effect depends a great deal on the strength of the solution. For fastest work, add oxalic acid crystals to water until no more will dissolve. This saturated solution can be diluted if you come to places which need less lightening. Oxalic acid crystals may remain on the wood after it dries, and require rinsing. Dry again, thoroughly, before you put on subsequent finishing materials.

Some of the special sealers function both as lighteners and as "sanding sealers." Thus, the sanding-sealer demonstrated here is closely parallel to the lacquer lightener on the opposite page. In addition, it contains subtle pigments which filter out light rays which cause wood to darken, and the wood maintains the depth of tone it has under the original finish, without darkening year after year.

Other bleaching materials include hydrogen peroxide and a product called Double-X, sold at paint stores. Double-X must be applied boiling hot for best effect. As a result it is not particularly appropriate for delicate projects, although it is widely used as a finish remover and bleach on old floors.

MAKING WOOD LIGHTER WITH STAINS. Although you normally think of staining as darkening the wood, it is easy to see that a stain lighter in color than the wood will lighten it. When this stain is white, the wood takes on a much lighter hue—and sometimes the side effects are interesting.

The white stain most often used is White Firzite or White Rez, sold at any paint store. The technique of handling these materials is exactly the same as for pigmented wiping stains. You brush them on—allow a few minutes for penetration—then wipe them off. Pigment clings to the open pores of the wood, blending with the natural color, and the "blonding effect" is quite pronounced.

White stains work best on coarse-grained woods—the same as other pigmented wiping stains. Their effect is great, also, on such materials as fir plywood, pine, and similar woods with considerable variation in density between spring wood and summer wood. They are least successful on the better grades of cabinet woods, and if you use them on such species as walnut and mahogany, take special pains to wipe all the stain off the surface of the wood, to prevent an undersirable veiling of wood tones. As with all pigmented stains, the effect is relatively slight on hard, close-grained woods such as maple, birch, and beech, unless you leave quite a bit of white on the surface.

Another method of using white stains is as a "glaze." This finishing technique involves a thin, paint-like material which you brush on, then wipe selectively, leaving much or little on the surface. Glazing is covered completely in Chapter 16.

Open grain woods are easy to stain white. The sample at left is treated with a white wiping stain, then varnished. Notice how the stain reversed the color pattern, making the darker (most porous) areas whiter — since they take more pigment. Greater lightening effect plus subdued grain differences show up (right) when you use white as a glaze. See text and chapter on glazed finishes.

White pigment in paste wood filler gives this result — lighter, but with emphasis on the pores. All three samples are oak plywood — the first two rotary cut, this one rift-sawn.

WHITE FILLER LIGHTENS WOOD. When the wood you are working with is oak, pecan, lauan, Philippine mahogany, ash, elm, or other open-pored species, you can produce an interesting look—and lighten the effective color quite a bit—by using a white wood filler. To make the filler, use white pigment in neutral paste. Keep the mixture fairly thick, and avoid using penetrants. Seal the wood with a washcoat of shellac at about 1-pound cut, to prevent whitening of the surface, since you want the pigmentation to be confined entirely to the filler in the pores. Wipe carefully, to avoid veiling of wood tones. (Complete details on using paste wood fillers are included in the next chapter.)

For a wood finish so light that it would have to be classified as a novelty, you can combine the bleaching and the paste filler.

SPECIAL FINISHES THAT LIGHTEN. There are certain special finishing materials, not widely enough known, which lighten the color of wood by *minimizing the degree to which regular finishes darken it.* These wood lighteners are often combined with sander-sealer properties. An example is McCloskey's Kwik-Sand. The material is varnish-like but contains solids such as calcium carbonate and silicates. These solids give the wood a light-colored overlay, they provide a good sanding surface, and they hold subsequent finishes from penetrating the wood and darkening it.

Since an oil-resin type of lightener-primer contains pigments, it helps hold the color of the wood so that it doesn't darken so greatly with age. For that reason it is an excellent material on paneling in the cherry-and-lighter wood colors, in situations where you want to hold the light color.

CHAPTER ELEVEN

HOW AND WHEN TO USE WOOD FILLER

No wood is entirely smooth, no matter how carefully sanded, no matter how fine-grained by nature. When the tree was growing, its trunk contained a great deal of water. When the log was sawed and planed and dried, the water disappeared, leaving spaces. In oak, chestnut, pecan, and similar woods, the holes are big. In holly, gum, and some others the pores are so small you can hardly detect them. But—holes there are, wherever saw or plane or sandpaper cut into a wood cell.

In some cases you couldn't care less; in fact, you like it that way. When you give wood a penetrating resin finish the pores stay open, and that's what makes the wood look as natural and untouched as it does.

In other cases, you don't want the wood pores to show under any conditions. If you are refinishing a Sheraton sideboard for instance, or any other high-style traditional mahogany, you want an absolutely glass-like surface. The same would be true of a rosewood guitar, which wouldn't look right to the eye of a real expert with anything less than a rubbed and polished surface so shiny—but not glossy!—he could see himself in it.

An obstacle to the absolutely smooth finish is that the thin film of varnish follows the hollows of the wood pores. Even after four or five coats, you can still see the pores in woods like oak and Philippine mahogany, although they pretty well disappear in maple, birch, beech, and other finer-grained species. To prevent this, use wood filler. It fills the holes. Your finish stays up on the level.

KINDS OF WOOD FILLER. The filler most commonly used is called "paste wood filler" for the reason that you buy it in a consistency about like peanut butter, although you thin it for use. This is the kind of filler used for open-pored, coarse-grained woods. It is composed of ground silicates and other solids which fill the pores up level. The finish coats ride across the filler—perfectly smooth.

The other kind of filler is much like a sanding sealer in that it is heavy-varnish consistency. It, too, contains solids, but they are usually finer and more transparent than those used for paste filler. The place for such fillers

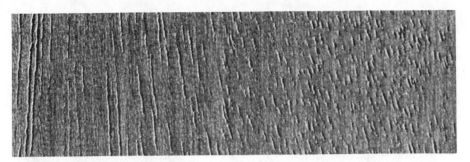

Close-up of a small area of mahogany, clean and bare, shows the textures of wood grain. These depressions are filled level in the wood-filling process.

The same area of mahogany, filled with a medium-dark filler and topcoated with two coats of varnish, rubbed. Notice that the wood grain still shows, but the surface is absolutely flat and level. Rubbed, it shows no glare, although light bounces brightly off the coin.

is on the finer-grained woods such as birch and maple. Many good finishers shy away from the sanding-sealer-fillers because they may give adhesion problems. Undoubtedly you have seen maple or birch furniture a few years old with a finish that seems to be "glued on" by some sort of a semi-transparent material and is coming unstuck. Nine times out of ten this "glue" is a brush-on filler. You find this condition most often when the filler is in a lacquer base—less often when it is a varnish or oil-resin base. To avoid it, many craftsmen simply brush on a coat of good varnish to act as a sanding sealer and get excellent adhesion.

HOW TO USE PASTE WOOD FILLER. If you have access to a fairly sophisticated finishing materials supply house, you can buy paste wood filler in all of the standard colors for regular finishes on mahogany, walnut, and other much-used cabinet woods. If your only convenient source is a regular paint store, nothing much is lost, since you can match the professionals' colors with ease, using "neutral" filler.

Paste wood filler is naturally a light cream color—the color of the ground solids wet with oil. You would rarely use it this color, except for a light finish on korina, prima vera, or other naturally pale wood. However, the color is a good starting place for tinting the filler any color you want, using universal colorants or pigments in oil. As with stains, the colors you'll use most often are Van Dyke brown, burnt umber, burnt sienna, and black —with white for special effects (see Chapter 10, on making wood lighter).

The material is too thick to use as you buy it and must be thinned with turpentine or paint thinner. Do not use oil to thin paste filler; it doesn't harden well enough or soon enough to give you the smooth surface you're after. When properly thinned, wood filler is about like cream—about as heavy as a good, rich wall paint. It should brush easily, but retain enough body to work into the pores of the wood and not lose much or any volume through evaporation of the thinner.

After you have thinned the filler, add the pigment. Generally speaking, unless you want some special effect, the filler should be darker than the *stained and topcoated wood.* The finish looks best if the pores of the wood show up darker than the rest of the surface. Some people like this difference to be great, and lean toward filler that is almost black. For a starter, however, try for a shade or two darker.

The color should match the wood—not necessarily precisely, but at least as close as the following rough formulas will provide:

Mahogany: Burnt umber. Sometimes a little rose pink, if the color is a little red. For brown mahogany, add a little black.

Walnut: Van Dyke brown. A little umber takes some of the warmth from Van Dyke.

Oak floors—clear: Just a touch of burnt umber. Too much pigment in the grain of floors makes them look dirty.

Oak floors—stained: Use the colors recommended for walnut.

Both burnt umber and Van Dyke brown are such universal woodhue colors that you can use them to darken filler for almost any wood and for almost any finish.

Some finishers use stain to tint the filler, if they are using oil or pigmented wiping stain. This practice is all right, but it is sometimes difficult to make the filler perform properly when it contains the liquids of the stain. *Remember:* the trick with wood filler is to brush on a liquid which will gradually turn back into a firm paste, so that it will stay in the wood pores.

Brushing the filler on. You scrub filler on, more than brush it on. Use a stiff brush. Swab it on with the grain first, making sure that coverage is complete, although uniformity is of no importance and neatness doesn't necessarily count. Then scrub across the grain. Your objective is to work the filler into the wood grain as deeply and as firmly as possible. You want to pack it in tight so it won't come out when you clean off the surplus.

In a few minutes, the filler will begin to dull over. This means that the liquids are evaporating, leaving mainly solids which are again in a paste form. There is a fairly exact point at which the drying filler is just right for

Best way to buy wood filler is in the natural (creamy) color. Then, as you need it, color working quantities with universal colorants. At the same time, thin the filler to heavy-cream consistency with turpentine or paint thinner.

Brush the filler on with every-which-way strokes, attempting to work it into the grain as well as possible. Neatness does not count, but coverage should be complete and the filler must be thick enough, in consistency and in application, to fill the pores.

In half an hour or so the brushed-on filler dulls over, indicating that the thinners have evaporated. Now you must remove the excess filler, taking great pains not to dig it out of the pores. An excellent method of starting this operation is by scraping the filler with a slightly slicing motion, using the edges of an old deck of cards. This action leaves the filler in the pores, but squeegees most of it off the surface.

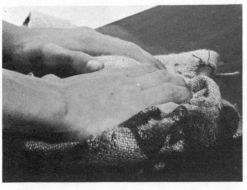

Finally, wipe the surface *crossgrain* with burlap or other equally coarse material. Do not use soft cloths which would work into the pores. Wipe as clean as possible. Then give the filler 24 hours to dry. If any remains on the surface — and vestiges may — touch it up very gently with 6/0 paper. Dust, and you are ready for topcoating.

wiping. With practice you'll discover this point. If you wipe too soon, you'll lift filler out of the pores. Wait too long and you'll find it difficult to get the material off the surface.

One good trick is to squeegee the filler cross-grain with a straight, true edge. If you have an old deck of playing cards around the house, they make perfect scrapers. You can use a wide putty knife, edges of wood scraps— anything that rides on the surface, scrapes off surplus filler, and tends to work the material firmly into the pores.

After this, switch to a coarse cloth, excelsior, burlap, or other heavy material. *Work cross-grain only.* Never wipe filler with the grain, or you'll surely wipe it out of the grainwise pores. Wipe as clean as you can. Any filler you leave on the surface obscures the grain and dulls the finish.

Give the filler overnight or longer to dry. If your wiping was thorough, you are now ready to apply the finish. If you can detect any roughness by feel, or if you can see any filler on the surface, get out some extra fine flint paper and a felt-covered sanding block and give the surface a gentle once-over. You want to cut away the unwanted filler, but never sand enough to remove any wood. If you do, you'll open up more wood pores which should be filled.

Industrial finishers use lacquer-base fillers under lacquer, both to gain the drying speed and to be sure the filler is compatible with the topcoating. Under ordinary circumstances, there is no need to worry about a lacquer finish softening ordinary wood filler if you give the filler twenty-four hours or longer to dry.

HOW TO USE LIQUID FILLERS. There is little difference between the liquid filler paint stores sell and the sanding sealer discussed in Chapter 5. Varnish or lacquer based, they contain a small quantity of solids, usually silicates. These solids fill the pores. Since liquid fillers contain relatively little solid matter, however, they must be used *only on fine-grained woods.* And, in the opinion of many good wood-finishing experts, there is no need to fill such fine-grained species as maple, birch, poplar, basswood, beech, cherry, gum, etc.

Brush on liquid filler about the same as you would a regular finishing coat. There is no point in an extra heavy application, as long as coverage is complete. When the filler is dry, sand carefully. You must get all of the filler off the surface of the wood, leaving it only in the pores, or it will cloud the finish and it may yellow with age. Go at the surplus dried filler with 3/0 or coarser paper, to get quick results. Then switch to 6/0 when you have removed most of it. Be careful not to sand below the original surface, or new pores will open up.

Shellac has a tendency to fill the pores of any wood that is not too coarse by nature. Used in two coats, with sanding between, shellac produces a good, level surface, over which protective coatings of varnish go on very smoothly. Remember, however, that some of the synthetic resin varnishes do not adhere well to shellac.

CHAPTER TWELVE

HOW TO GET THE MOST OUT OF SHELLAC

OF THE three most common on-the-surface finishes for wood—varnish, shellac, and lacquer—shellac is by far the easiest for most people to use. The reason for this is the way you apply shellac, in addition to its unique characteristics.

Almost any place inside the house where you want a clear finish, shellac is a candidate. It puts a beautiful finish on furniture. It is much used for floors, where its flexibility is a good characteristic and where the quick-dry of shellac makes it possible to do an entire floor job in one day. The only drawback with shellac is a low resistance to water damage, along with a tendency to soften under alcohol and other chemicals. For that reason, shellac is not often a good choice for projects which may be subjected to such rigors.

Shellac is never a one-coat finish. You *build* a finish with it in two or more coats. With each coat you have a renewed opportunity to repair any mistakes you made in the preceding coat, and to smooth out the blemishes caused by dust. While it is true that you can do this with varnish, varnish dries so slowly that a three-coat job takes three days. You can do it with lacquer, too, but lacquer dries fast and many now-and-then finishers find it hard to apply smoothly. Shellac, meanwhile, dries dust-free in a half hour or less. You can recoat it in three or four hours. Used properly thinned, it flows on easily and hardly ever shows lap marks, since the brush-over tends to soften the preceding brush stroke blending the two together.

Few finishes have greater natural adhesion, and since successive coats tend to fuse together, most people end up with a good, smooth, level surface that is well adhered to the wood. Shellac is flexible and it does not check or craze provided it is properly applied in several thin coats.

Shellac presents you with a choice of surfaces. It builds to a good sheen, rubs to a good satin, steel-wools to a flat finish, takes waxing very well. (In fact, waxing is considered essential with shellac on any surface subject to abuse, to add water-resistance to the finish.)

To many an eye, nothing beats shellac for the color it gives wood— warm and woodsy in a depth about halfway between the cool color of lacquer and the dark warmth of penetrating resin sealers. The tone shellac gives to dark woods is one of the reasons why many good finishers brush on a first coat of shellac, even though they plan to gain the greater durability of varnish with subsequent coats. Never use a urethane varnish over shellac, however; the adhesion is very poor.

Shellac is made from a natural resin secreted by an insect, dissolved in denatured alcohol. It is about as economical a product for wood finishing as can be made. In today's age of sophisticated paint chemistry, it is one of the last remaining unsynthesized materials for wood finishing.

HOW TO BUY SHELLAC. While it is practical to buy larger and therefore more economical put-ups of varnish, enamel, paint, etc., you must never buy more shellac than you need for the immediate job. Shellac is not stable in storage, and about six months after you buy it, chemical changes occur which may affect drying and durability.

In practice, you can test any shellac you happen to have left over. If it dries, you have nothing to worry about. But, if it dries slowly or stays tacky, throw it out. Never make this test on an actual project, because removing the nondrying shellac is a nuisance. Brush a few strokes on a scrap.

Never make the mistake of adding some more alcohol to an out-lived, slow-drying shellac, hoping to make it dry. All you'll get is a thinner shellac that won't dry.

Some shellac manufacturers date their cans. You can estimate whether you'll be able to use up the can before the expiration date. Other manufacturers just let you guess how long the stuff may have been on the dealer's shelves.

Don't try to save money buying off-brands of shellac. It doesn't cost enough to permit a significant saving, and you run the danger of buying an inferior product, one that is out-dated, or one in a nonlined metal can which is pretty sure to turn black because of the chemical reaction between the shellac and ferrous metals.

Every time you buy shellac, buy at least an equal quantity of alcohol. You almost never apply shellac as thick as it comes from the can. Most finishers look at it as a stock solution, which you mix with additional alcohol at the time of use.

The meaning of "cut". Shellac is designated as "4-pound cut," "5-pound cut," etc. This means that the resin is dissolved in alcohol in the proportion of 4 pounds of resin to a gallon of alcohol. The 3-pound can be used as is on floors, but it, like other concentrations, must be thinned for fine work. Which cut you buy makes no difference, since you'll thin it further, anyway.

White vs. orange shellac. By far more white shellac is used than orange, although the amber cast of the orange is preferred often for use on walnut, mahogany, and other dark woods.

HOW TO USE SHELLAC. Here is a table of dilutions for shellac:

BASIC CUT	DESIRED CUT	ALCOHOL PER QUART
5-pound	3-pound	⅞ pint
5-pound	2-pound	1 quart
5-pound	1-pound	⅔ gallon
4-pound	3-pound	½ pint
4-pound	2-pound	¾ quart
4-pound	1-pound	2 quarts
3-pound	2-pound	½ pint
3-pound	1-pound	3 pints

In actual practice, it is not critical for your mixture to be a precise, round number, and the table is mainly for convenience; it is simpler to say "use a 1-pound cut" than it is to say, "if you have a 5-pound cut, add alcohol at the rate of two-thirds of a gallon per quart."

Experience will soon teach you that a dollop of shellac and two dollops of alcohol give you about what you want. However, it is good craftsmanship to know how you produced any finish, so that you can repeat the performance if it's worth it. For that reason, most finishers dilute their shellac quite accurately at so many pounds.

Within reason, the thinner you use shellac, the better. At about 1-pound, shellac brushes on fast, and tolerates a great deal of ineptness. With so much alcohol in it, it dries quickly for recoating. Besides, it penetrates and adheres better. At thicker cuts, adhesion and penetration are not so good, the film doesn't dry as hard—and drying is quite slow.

It usually takes the first coat of shellac about an hour to dry hard enough for between-coats sanding. In good conditions, half an hour may be enough. In muggy weather, allow more time. Sand lightly. Brush on the second coat. Since there is no penetration with this coat, drying to sandable hardness may take two or three hours. Again, sand to level the surface. Coat again, for a high-grade finish—and perhaps again, if you want more build.

Many operators put three coats on surfaces such as the sides of a chest, then put one or two extra on the top. This is good practice, as long as the film is entire on the sides and other noncritical areas, so that the color uniformity is good.

Below is the most foolproof way to use shellac; as your skill improves, you can work with 2-pound or 3-pound and use fewer coats.

Put it on wet. Shellac forms the smoothest coating when you dip a full brush and flow the material on wet. Brush slowly, so you don't create bubbles. Lap each brush stroke over the last one. Work for uniformity of coverage, although the multicoat system recommended usually results in good evenness without more than normal attention to smooth brushing.

Sand between coats. The first coat of shellac, over the best-sanded wood, produces a tooth much like raised grain. Sand it off smooth. Shellac always clogs the paper quickly, since it gums up under the friction of sanding. Either use cheap flint paper and lots of it—or an open-coat paper that will absorb more shellac before it becomes useless.

The second coat may show some of the same tooth—and perhaps the little sharp pinnacles caused by dust. Again, sand, knocking off the dust nibs and leveling the surface. Be sure to use a sanding block, so that you work down the high spots, rather than following the ups and downs as you would without a block.

When you have built up the required thickness of film, take special pains to sand level. Then switch to 4/0 grade steel wool. Working carefully, watching the small remaining glossy areas, rub grainwise with the steel wool until no shiny spots are left.

At this point, the shellac film is perfect for waxing. But—wait twenty-four hours before you wax, so the shellac can harden. Use a good quality paste wax. A good way to apply it is with a small pad of steel wool. Buff the hardened wax with lamb's wool on a rotary sander pad. Or, use soft rags. Two or three coats may be needed for a top-grade luster.

Less critical work—such as paneling, trim, etc.—can be handled with two coats of shellac at about 3-pound cut. Sanding the final coat is optional. Shellac's natural gloss is usually acceptable for such applications, about halfway between the high gloss of a spar varnish and the satin sheen of a semiglossy varnish.

Cleaning brushes. Anyone should be delighted over the fact that shellac brushes rinse clean in a mild solution of household ammonia and water. This economy of clean-up is one reason why it is efficient to use a roller when applying shellac to large areas.

Ingredient for the much-admired finish called "French polish" are shown here. As the text explains, you put the finish on with a pad of cloth, starting with shellac, then add oil later. You can add oil to the shellac — or you can dip the pad first in oil — lightly — then in shellac.

THE ART OF FRENCH POLISHING. One of the loveliest finishes for fine furniture, called French polish, is done with shellac and a clean, lintless rag. The surface must be carefully sanded and stained with a water stain or a non-grain-raising material.

Pour some shellac at about 1-pound cut into a saucer or pan. Dip the rag in the shellac and apply it with smooth, quick, even strokes, grainwise. When this application is dry, hit the surface lightly with 6/0 paper and repeat the wipe-on process. Continue until the film is thick enough to show a slight sheen.

At this point, put several drops of boiled linseed oil in the shellac mixture, and change the application technique from one of wiping on to one of polishing in a circular movement. As more and more coats go on, add more drops of oil.

When you get to a degree of build that satisfies, stop. This finish takes time, but owing to the polish-on technique, it is extremely smooth and lustrous—worth all the work.

CHAPTER THIRTEEN

VARNISH–THE MOST USEFUL CLEAR FINISH

THE BEST all-around clear finish in the hands of the typical wood finisher is varnish. It is more difficult to use than shellac; but it is far more durable in virtually any situation, and the problems of producing a fine varnish finish are easily solved with ordinary patience and willingness. It is slower drying than lacquer, but again durability favors varnish in most applications where speed of drying is not a critical factor. Varnish is truly the universal finish, giving superb performance on any wood surface in any situation. Furniture, floors, paneling, trim, cabinets—anywhere in the house, varnish is a good choice. Outside, varnish has shortcomings since it fails much more quickly than paint or enamel, owing to the effects of the weather and sunlight on its film.

WHAT VARNISH IS. The day of the natural resin-and-oil varnish is gone. Man-made resins have taken over just about entirely. Such ingredients as linseed oil are replaced with man-made solvents. Whereas old-timers selected (or modified) their varnishes to meet dozens of different situations and needs, the entire range of requirements is met today by three or four varnish variations you can find on the shelves of any paint store.

By far the most interesting (and recent) of these is *water-thinned* varnish, a product that is part of the finishing-chemistry industry's trend toward materials that are easier to use and safer to use. In oversimplified terms, the industry has added chemicals to the formulas that make those tough resins soluble in—mixable with—water, instead of in petroleum-derived thinners. Once they have dried, however, they are just as tough as their petroleum-thinned cousins.

The major advantage is the way brushes wash clean with soap and water —sometimes even more quickly and easily than we have come to expect with latex paints. Another feature you'll discover in some of the water-thinned clear finishes—varnishes—is the way they go on.

The flow-out is good. The application has a milky appearance as it goes on, and you can judge the uniformity of coverage by the degree of milkiness. As it dries, it turns clear.

The drying time is fast, making it possible to put on two or three coats in a single day, as opposed to the coat-a-day schedule most often approved for regular varnish.

Give the water-thinned material a try. You may be willing to overlook their tendency to raise the grain in view of the fact that they give wood all the beauty you get from oil-based finishes.

The resins used in today's varnishes are usually alkyd, phenolic, or urethane. The alkyds tend to be a little less expensive. They come in glossy and satin. The phenolics are usually found in exterior varnishes and some of the special clears for marine brightwork, although they also appear in varnishes intended for interior work. They are usually glossy only. The urethanes are still more expensive. Their wear resistance and ability to hold up under chemical abuse is greatest. Glossy, satin, and flat are available in the urethanes.

There are, in addition to these basic varnishes, several materials with entirely different characteristics. One of these is a vinyl formula which has quick-dry characteristics similar to shellac and lacquer, such as Haeuser's Quick-15. Rez 20 is another fast-dry clear. These products are resistant to water and alcohol, they are excellent in flexibility, and dry to a good lustre. For woodwork, trim, paneling etc., these quick-dry materials are an excellent choice. However, if you want to rub up a high finish on a good piece of furniture, an alkyd or urethane is better. (Read labels carefully on these vinyl and other recently introduced varnishes, since instructions frequently run counter to much common practice with regular varnishes.)

Some excellent varnishes combine alkyd and vinyl, with advantages in flexibility over straight alkyd.

WHICH TYPE OF VARNISH IS BEST? In the end, you'll decide this for yourself. Performance is close to the same with all of them. The urethanes brush on very, very easily, the alkyds only slightly less so. Urethane drying time is shorter than it is for alkyds or phenolics—but not as short as for the vinyls. Vinyls are the most flexible, thus resistant to checking or crazing when the wood swells and shrinks—but only a little bit more so than the urethanes. Urethanes rub well with a minimum of clogging of the paper—but so do alkyds.

The accompanying chart will act as a guide, but you will never know which varnish you like best until you've given several types a good and thorough trial. And the trial will never be conclusive unless you know the advantages and disadvantages of shellac (Chapter 12) and the tremendous simplicity and quality of the penetrating finishes (Chapter 7).

It is difficult—even impossible—for the average person to judge the quality of varnish in advance of use, for it is measured largely in *whiteness*, in resistance to *yellowing*, and thickness of film. Phenolic varnishes are usually the most yellow, and tend to turn more yellow. Urethanes and vinyls are the clearest and show very little color change. Alkyds vary quite

Basic Characteristics of Varnishes

Kind of varnish	Sheen	Color effect	Protection	Number of coats	Drying time	How to apply
Glossy	High shine	Darkens wood. Pale to quite yellow film.	Excellent	2 or 3	24 hours	Brush or spray
Semi-gloss	Low shine	Same	Good	1 or 2. If more are needed, start with gloss	24 hours	Brush or spray
Satin	Lustre	Same	Good	Same	24 hours	Brush or spray
Flat	Low lustre	Same	Fair	Same	24 hours	Brush or spray
Urethane	Available in gloss, semi-gloss satin, and flat	Darkens wood. Film is quite pale.	Superior in gloss. Excellent in dulls.	Same	Varies; see labels	Brush or spray
Vinyl	Gloss or semi	Darkens wood less. Film is quite pale.	Good except abrasion	Same	15 minutes. See labels	Brush or spray
Oil	None	Darkens wood	Poor	Many. See text	See text	Brush or swab

a bit, and your best assurance of quality is a well-known manufacturer. This, of course, is true of any varnish you buy. A penny saved is far from a penny earned, if the saving costs you extra work—or a bad job.

PREPARING THE SURFACE FOR VARNISH. Since the wood shows through varnish—the more it shows the better—it must be as nearly perfect as possible before you put on the first coat. In Chapters 3, 4, and 5 removal, clean-up, and sanding are discussed fully.

Bare wood. Examine the project thoroughly for dirt, finger marks, streaks caused by abrasion, and other blemishes. Clean them off with mineral spirits, lacquer thinner, or a multisolvent product such as Cleanwoode. If necessary, sand, but if you sand, use the finest paper and run over the entire area which is blemished. Otherwise, the freshly sanded area may not take the coloration from the varnish evenly.

Stained wood. Be sure stains have had adequate drying time. Give the stain job a last minute inspection, to see if there are any places where they are too dark and might produce muddy areas. Lighten these areas carefully, using fine steel wool on pigmented wiping stains and the proper solvent on water stains or non-grain-raising stains—i.e., water or NGR thinner. Do not forget that varnish will change the color of the stained wood. As the chapter on stains explained, you should make tests in advance of stain-plus-varnish to determine the exact shade. If you have not already done this, you may want to put a brush-stroke of varnish in an inconspicuous spot as a last minute check. If the color is not right, you can darken it slightly with a reapplication of diluted stain or lighten it as mentioned above. If you add more stain, be sure to wait the proper drying time.

Old finishes. When an old finish is in good enough shape so that it needs nothing more than a rejuvenating coat of new varnish, make sure that it is absolutely clean. Wash it with detergent and water and rinse, or mop it with mineral spirits, lacquer thinner, or a product such as Liquid Sandpaper or Wilbond. All oil, wax, grease, and other contamination must be removed. If the surface is glossy, scarify it with fine sandpaper, to insure good adhesion.

DUST IS THE ENEMY OF VARNISH. No matter how carefully you work—unless it is in an air-filtered, humidity-controlled, sanitary room— you'll have trouble with dust. What you must do is eliminate as much of it as possible, then learn what to do about that which, inevitably, remains.

To begin with, dust is the biggest headache when you use a glossy varnish and leave the final coat glossy—that is, unrubbed. Dust *on* the surface or which *settles on* the freshly applied varnish forms tiny pinnacles as the varnish accumulates around the dust speck. If there are only a few such specks, you can take care of them with a picking stick or a small artist's brush. If there are many, the finish that is intended to be slick and smooth turns out to be toothy and rough.

A little dust is less of a problem with rubbed finishes, since the rubbing eliminates the tiny pinnacles. This is one reason why many finishers consider a rubbed finish easier to do than a truly fine glossy finish.

Here are the steps toward cutting down the dust problem:

1. Do your varnishing outside the workshop. Pick a room where there is little traffic. (How can you beat a spare bedroom?) Put the piece to be varnished in the room, and wait an hour or so for the dust raised when you brought it in to settle. If the project can't be moved—as for instance, paneling—dust the room and vacuum the floor. Then stay out of the room for a couple of hours. When you go back in, try to raise as little dust as possible.

2. If you must do the job in the workshop, do a thorough clean-up before you begin. Sweep the floor with wet sawdust as a sweeping compound. Again, give the dust time to settle. Take the project out of the shop while all this is going on, then bring it back in.

3. Wear clothes which do not shed lint and dust. Generally dacron and other synthetic fabrics produce less dust than cotton. Wool is particularly bad.

4. Dust the project by using an old, clean paint brush. Follow this with blasts of air from a sprayer compressor. A tire pump works well too. You can even get the job done blowing through a soda straw. Pay special attention to cracks and carvings which always accumulate dust which the brush brings out and spreads about.

5. Finally, use a tack rag. Dust which defies all other forms of removal responds to a tack rag.

How to make a tack rag. The quickest and easiest way to get a tack rag is to buy one at a paint store. It is nothing more than a piece of cloth impregnated with varnish which becomes tacky (hence the name) so that dust sticks to it. Make one this way:

Take a piece of cotton cloth about the size of a dish towel or smaller. Dampen it thoroughly with water. Wring it as dry as you can. Wad it up and pour an ounce or two of turpentine over it. Work the wad in your hands, distributing the turpentine throughout the cloth. Wring it dry again. Ball the cloth up into a wad once more and pour an ounce or two of varnish over it. Thoroughly massage the wad to spread the varnish. Open it up and examine the color. It should be uniformly yellow. If it isn't, work the wad some more.

The tack rag should be *just barely damp.* Hang it up for a few minutes to dry if it is too wet. As you use the rag, it will become too dry. Add small amounts of water and turpentine to keep it moist.

Using a tack rag is a simple wiping operation. Just fold it to size and wipe the surface with it. Between uses, keep the rag in a glass jar with a screw-on lid, or put it in a sturdy plastic bag and twist the opening tight.

APPLICATION TECHNIQUES FOR VARNISH. The first coat of varnish on bare wood or wood colored with water or NGR stains should be thinned with about 1 part of turpentine to 4 parts of varnish. Brush this coat out well, using it as a sealer. *When you seal with the same varnish as you'll use for the rest of the job, you get better results. You do not run the risk of different characteristics fighting each other on the wood.* For instance,

Always try to varnish with a white wall or a window in back of the work, so that you can watch the glare — the shiny area in this photo. It will tell you quickly if you are over-applying, if you are skipping — and if you are picking up dust. The tiny dark specks on the varnished area shown here are dust specks which you could pick up with a picking stick — see text.

if you seal with a material which has a different degree of flexibility, you set up stresses between the sealer and the varnish, which will eventually cause the finish to fail. Many of the best varnish practitioners, working for a perfect job, even use thinned varnish as a sanding base for the final smoothing.

Over a pigmented wiping stain, which already contains resins which seal, you can use varnish without thinning.

Second and all subsequent coats should be used as they come from the can unless you find that there is too much drag on the brush. If this is the case, use a very small amount of turpentine. Stir slowly—to avoid bubbles—but thoroughly. If possible, let the varnish with turpentine stirred in stand for a few hours, to homogenize.

These coats of varnish should be full, but not puddled. With a little experimentation and practice you will learn to put on the exact amount of varnish *which will level*. If you use too little, you'll see skimpy spots where there is too little varnish to flow out smoothly. If you use too much —more than is needed for leveling—drying time will be overlong. And, you run the risk of crazing and checking later on.

How to brush varnish smoothly. Whenever possible, work on horizontal surfaces. Disassemble a piece if you can—removing drawers, doors, etc., so they can be laid flat. Tack up a piece of white paper or a white sheet, or work in front of a white wall, so that you can easily observe the glare produced by well-leveled varnish.

Give the surface one last swipe with the tack rag, then start at the far edge and work toward you. Work out any method of laying the varnish on that is easiest. Most professionals lay on two or three brush-wide stripes with about a brush-width between them. Then they brush crosswise to

This is a varnish cup with a "strike wire" made from a tin can. The wire is a piece of coathanger stuck through two nail-punched holes in the can and bent over. When you gently wipe — or strike — the excess from a brush, it drips into the can cleanly, without over-edge dribble or excessive foaming.

smooth out. Finally they tip off the varnish using just the ends of the bristles. Brush handling is illustrated in the accompanying photographs. The standard advice is to dip only one-third of the bristles into the varnish. Violate this rule if you want to; dip less, dip more, depending on whether you are going fast over a large surface or working slowly over a small area or around carvings. When you first dip the brush, massage it for a few seconds against the side of the can, to work the varnish well into the bristles.

Continue laying on varnish, smoothing, then tipping until the entire surface is covered. Generally, it is best to work from end to end of an area that is longer than it is wide rather than from edge to edge. As you complete each cycle of lay-on, smooth-out, tip-off, make a final leveling tipping stroke from the new varnish into the varnish just applied. Observe this operation carefully, and watch for any lack of smoothness. If the joining strip doesn't level out in a minute or two, it is a sign that you should go through the three-stage cycle over a smaller area, since the edge is beginning to set a little before you get back to it. With practice you'll learn just how large an area you can cover and still show no lapmarks.

Do not slap-slap varnish on. Stroke smoothly and slowly, or you may whip up bubbles. Sometimes bubbles break quickly and the varnish levels. Too often, however, bubbles break after the varnish starts to set, and the result is a small depression which is difficult to sand out.

Proper brush angle for varnish — or any other finish — is about 45 degrees, so that the "chisel" of the brush meets the surface and puts the maximum amount of bristle in contact with the wood. The proper degree of bend in the bristles is shown here, too.

Avoid what you see in this photo — the brush falling over the edge of a surface. It produces a "fat edge" because the edge squeegees extra varnish off the brush. Drying is slow, rubbing is difficult, and the result is as much of a woodfinishing disgrace as is a sag.

Use the tack rag, wiping the unvarnished surface as you go along. It will save lots of work later on.

Special problems of curved surfaces. Carvings, moldings, and other surfaces which are not flat present problems in that the irregularities tend to squeegee varnish off the brush in greater quantities than necessary, causing sags and puddling. For that reason, work with a fairly dry brush on irregular surfaces.

Rounds and turnings. The best method of handling round legs, spindles, stretchers, and the like is to apply the varnish by brushing back and forth across the round. Then, level with lengthwise strokes. Usually, when spindles are small, you work better with a smaller brush than you'd use on flat surfaces.

Moldings. Regular moldings, such as ogee, cove and bead, etc., which are not carved (as are egg-and-dart and others) are simple to handle with lengthwise strokes. If you brush across moldings, you usually fill the depressions and skimp the high ridges.

Carvings. Putting varnish on intricate carvings is more lifting than brushing. First, spread the varnish over the carved area. Dab it well into crevices and deep depressions. Then, by placing the brush in carvings and lifting it out with a gentle brushing motion, you gradually dry up the over-wet areas. Strike the excess off your brush by pulling it across the strike wire in your varnish can. If dust has accumulated in the carvings, the varnish you pull out will be filled with it. Dispose of this dusty varnish by striking it off on the edge of a tin can, which you can then throw away. (Use such worthless varnish to make tack rags, picking sticks.)

True panels. The proper method of applying finish to a true panel— for instance, a paneled door—is to do the panel first, then the frame. As you coat the broad surface, catch the inner edge of the frame. Then pull out the excess varnish gathered in the crack, by putting the tips of the bristles in the crack and brushing gently away. Finally do the frame and the edges.

Corners and edges. If you let a brush stroke go over an edge or an outside corner, you will leave sags on vertical surfaces and what finishers call a "fat edge" on horizontal surfaces. To avoid this, all smoothing and tipping strokes *must start on the flat* and go toward the edge. *Precisely at the edge,* lift the brush so that the bristles take off before they can flip over the edge and leave an extra-thick deposit of varnish. If you start the tipping strokes in the air, gently lowering the brush while it is in motion, you won't leave starting marks on the flat. Master this lift-off at the edges and you'll never be guilty of leaving a sag or a fat edge.

Inside corners—such as the point where shelves meet uprights—should be pulled out, to prevent a thick fillet of varnish.

How to use a picking stick. A picking stick is a splinter of wood with a small glob of sticky rosin on the end. It looks something like a wooden match; the rosin is the head. To make one, first dissolve a small amount of powdered rosin in varnish. (Buy a rosin bag at a sporting goods store or a chunk of violin bow rosin at a music store and crush it.) The mixture should be about as thick as cold honey. Dip the tip of the stick in it, moisten your fingers, and roll the glob of rosin into a ball. Dip it again, and roll again. This is a picking stick. Its gummy end will pick up a speck of dust.

Sighting at a low angle, so that dust specks are easy to see, touch the stick carefully to each tiny dust-caused pinnacle. Roll the dust speck into the glob with moistened fingers. Be careful not to press the stick into the finish, or you may make a crater which will not level again. Normally, however, the tiny depression made when you lift out the varnish speck will fill and level.

You can do a minor amount of dust picking with a small artist's brush or even with the split end of a wooden splinter or a broken toothpick, but a regular picking stick is worth the bother of making; it works much better.

Loose bristles. When a bristle loosens from the brush and lies in the varnish, pick it up by jabbing at one end of it gently with the tips of the bristles. It will lodge between bristles and you can pick it off with your fingers. A quick brush stroke levels the varnish where you picked up the bristle.

WHAT TO DO BETWEEN COATS OF VARNISH. To begin with, once a coat of varnish is on, never attempt to go back over it and fix up any bare spots where you missed or heavy spots where you lapped or left sags or fat edges. Catch them on the next coat, after you've used sandpaper to make them as level as possible.

Give varnish at least twenty-four hours to dry and harden between coats—if conditions are good. In muggy, damp weather allow thirty-six hours or longer. Experts say that the longer you let varnish dry before you recoat it, within reason, the better. The reason is the way modern varnish resins continue to "cure" after they have dried and hardened. If you overcoat them too soon, you rob them of part of their curing time.

An equally important reason for giving varnish plenty of time to dry is to make it easier to scuff-sand it for a smoother surface.

Special variations. Some varnishes—*read labels carefully*—utilize resins which call for different drying specifications. Typically, they can be recoated after four hours and up to six. If you cannot recoat during that period, you must wait twenty-four hours. There is a slight softening and bonding action between the two coats if the first coat is caught soon enough. Otherwise, it hardens too much for ideal adhesion; it must be left for twenty-four hours, then scuff-sanded.

Sanding between coats. Except for the special conditions noted in the preceding paragraph, you virtually never put on a second coat of varnish without sanding the first coat. How much you sand depends on the quality of finish you want. *Always* sand enough to scarify the first coat, to insure good adhesion. In addition, sand enough to level roughly applied varnish and to knock off the "whiskers" of any raised grain as well as dust pinnacles. Since you want to plane the surface with this sort of sanding, always use a sanding block. (See Chapter 5, on the use of sandpaper.) The ultimate in between-coat sanding is covered below under the "piano finish."

HOW TO USE SEMIGLOSS AND FLAT VARNISHES. For generations there have been semigloss varnishes. Originally they were made by stirring ground silica into the varnish. The silica broke up the glossy surface and reduced the reflectivity of the material. There was enough of this deglossing medium in the varnish so that an inch or more of it settled in the bottom of the can and had to be stirred almost constantly. The early less-than-glossy varnishes lost some of their durability due to the silica, and were never recommended for outdoor use or for hard-wear areas indoors. There was a slight cloudiness, caused by the gloss killing ingredients, and you were never supposed to use more than one coat of semigloss for fear of obscuring the wood. You built your finish with glossy varnish, then dulled it with a final coating of semigloss.

Modern semigloss, satin, and even flat varnishes are much different. They lose their gloss due to various chemicals which do not hinder durability. They are not as clear as glossy, but the difference is so slight that you can use two coats without harming the finish. This is, of course, a convenience, since you work out of one can instead of two. In addition, the semigloss materials scuff-sand particularly easily, giving them somewhat the character of sanding sealers when they are used as first coats in a two-coat system. If, however, you plan a three-coat finish, you may lose transparency if you try all three applications in satin-finish or flat.

The basic purpose of the less-than-glossy varnishes is to provide the richer satin look without the work of sanding and waxing the final coat. For this reason, they are excellent materials for paneling and other big-scale work. However, they give a lovely finish to good furniture for people who do not want to bother with rubbing. Two coats of a good satinsheen varnish on cherry or other fine-grain wood looks great, and a man on a galloping horse would find it hard to distinguish between it and a good rubbed finish.

THE CLASSIC 'PIANO FINISH'. The piano finish (so called because for many years the only object in the house with a really fine finish on it was the piano in the parlor) is the result of several coats of varnish, each sanded smooth, built into a tough, smooth, deep finish of excellent transparency and just enough luster to be in perfect taste. You can build the finish from any *clear* varnish, the whiter the better. If you use a urethane, the build is slower because each coat of varnish is thinner; the durability is the best, however. If you use a spar varnish you will get there sooner, because the film is thicker. However, varnishes labeled "spar" are often a bit yellow.

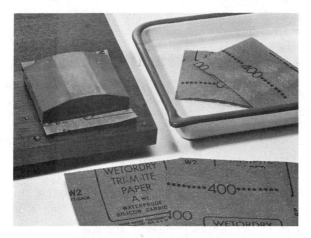

Fastest way to a rubbed finish is with waterproof sandpaper, 400 grit, used with water and a rubber block. Keep a few pieces of paper in a pan, to insure soft backing. Be sure there is plenty of water under the block.

These are the steps to a piano finish—the finest of all rubbed finishes:

1. Follow all the rules for preparing and smoothing the wood. Dust carefully.

2. Thin a small quantity of varnish with about one part in four of turpentine. Use this as a sealer. Sand very carefully, to restore all the smoothness of the wood. The sealer may raise whiskers of wood fiber. These must be sanded back to absolute plane. Use a sanding block. If the wood is open-pore (mahogany, walnut, etc.), apply wood filler. (See Chapter 11.)

3. Apply a second coat of varnish at can consistency. Do not thin unless the material is so heavy it drags too much on the brush. This is not likely to happen except with spar.

4. Using a sanding block, work over the surface until *all gloss is removed.* This may not be entirely possible with early coats, if there are depressions in the wood. Examine the results of the sanding carefully. A few tiny glossy spots in these depressions are a sign that you are planing well. In many cases, you may sand nearly all of the second coat off.

5. Take some 4/0 steel wool to the remaining glossy spots, and scarify them just enough to insure good adhesion.

6. Apply the next coat. Again, sand with a flat block. This time there should be fewer little glossy spots. Hit them with steel wool.

7. Repeat until there are no longer any—or only extremely minor—shiny spots. By this time you should have built up a good film thickness.

8. Now switch your rubbing to pumice and oil. Use raw linseed oil (motor oil will work)and powdered pumice sold at paint stores. Sprinkle the surface with oil and smear it around with your hand until the surface is entirely coated. Put some of the pumice in a big salt shaker—or improvise a shaker by punching nail holes in the pumice box. Sprinkle the pumice over the oily surface. Use a felt pad, and rub with the grain, applying only light pressure. Now and then wipe the surface dry to observe progress. When you have finished with the pumice, the surface should

Get a good rubber block by trimming one paint stores sell that is too big and cumbersome. Bandsaw makes the job easy, but you can do it with a handsaw, or a kitchen knife.

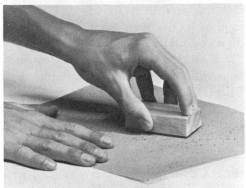

Another good wet-sanding block is a piece of pine with a chunk of old inner tube contact-cemented to it. Since the inner tube is not a uniform thickness of rubber, condition the face of the block by rubbing it over a sheet of sandpaper until it is uniformly dull. Be sure the sandpaper is on a flat surface.

These are the ingredients of a pumic-and-oil rubbing job. Use raw — not boiled — linseed oil, crude oil, or 3-in-One. Work down to plane with pumice, then switch to rottenstone for the finest, smoothest satin-sheen surface. (Alternative to rottenstone is automotive rubbing compound.) Block shown is two thicknesses of felt wrapped around a wooden block. You can buy regular rubbing felt at some paint stores.

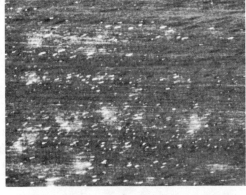

The shiny spots shown on this wood surface are guides to the amount of rubbing necessary. They represent low spots in a first coat after it is rubbed. After another coat of varnish is rubbed, most of them will disappear. After the third coat is rubbed, nearly all should be gone. The job is absolutely flat and smooth when there are no more spots.

be satin smooth. At this point, the varnish is smooth enough for most people, and is ready for step 10. For a smoother job, perform step 9.

9. As a final rubbing, turn to rottenstone and oil, handled the same as the pumice and oil in step 8. Rottenstone is finer than pumice, and it will return the varnish *almost* to a glossy finish but with none of the garish look unrubbed varnish has. It is perfectly smooth and transparent but without glare. Many finishers leave it that way, after taking pains to remove all oil and pumice with rags. Others go on to step 10.

10. The final step is paste-waxing. Smooth on a thin layer of a good wax—floor wax is hard to beat—with a damp pad of cotton cloth. Wait until this wax is dry, then buff it. If you have an electric drill with a disk-sanding attachment, use it with a lamb's-wool bonnet. Apply a second coat of wax, and buff again.

That is a piano finish—the loveliest on-the-surface finish you can put on wood—the working companion of the in-the-wood finish covered in Chapter 7.

RUBBING TIPS AND SHORTCUTS. The fastest and easiest method of sanding a coat of varnish smooth is with No. 400 waterproof paper, such as Wet-or-Dry. It is also the cheapest, despite the higher cost of the paper, since it doesn't clog when it is used with water, and it wears a long time.

The best sort of a block to use with wet sandpaper is rubber. Paint stores sell rubber blocks—which are a nuisance to use. However, you can cut them down so they are just a plain block of rubber, and they work well. Another way to get a rubber sanding block is to use contact cement to fasten a piece of innertube to a wooden block.

Sprinkle water on the surface—and sand. Keep a piece of paper in a saucer of water, so the backing will soften, and rotate the paper as its edges and backing dry. Many workers prefer to dip the paper in the saucer of water, letting it drain thus washing the accumulation of varnish powder off the paper. Now and then you hear recommendations to use soapy water; this does no good, and actually slows the job down. Plain water gives you the fastest and cleanest cut.

Depending on how fine waterproof sandpaper you can buy, you may be able to do an entire rubbed finish without using pumice or rottenstone. The equivalent to rottenstone is crocus cloth.

Rubbing carvings. Most finishers don't attempt to rub carved and coped surfaces as carefully as they do flat or mildly curved surfaces. For one thing, such a high finish is not necessary, because carved surfaces carry their own decorative weight.

Wet sandpaper—in fact any sandpaper—is useless on carvings, although it can be used on some simple moldings. Try steel wool for early coats. Then, toward the end of the schedule, use a small scrub brush to apply pumice and oil. Take special pains to clean the pumice out of small cracks and crevices. Any that you leave will dry white and give the carvings a scurvy look.

Maintenance. A piano finish carefully applied, using good quality varnish, should last a lifetime without any maintenance. If you wax it, you may have to maintain the wax, but when the finish is bare, all you must do is keep it clean. Now and then wash it with mild soap and water.

THE CLASSIC OIL FINISH. Although it is not a varnish finish, the classic rubbed oil finish must be included in any discussion of clear finishes. It is an easy job—about equal in labor to the penetrating resin finish— to which it is entirely inferior in every respect.

The material is boiled linseed oil. The technique is brush-on (or swab-on), allow penetration time, then wipe. You do this once a day for a week, once a week for a month, once a month for a year, and once a year for the rest of your life.

The oil finish gives wood a good, natural look and is particularly good on walnut. It is fairly resistant to heat, but has low tolerance for any other kind of abuse. One decided advantage is patchability. If you happen to scratch the finish, merely dab on some more oil and rub it in.

CHAPTER FOURTEEN

LACQUERS FOR BRUSHING AND SPRAYING

THERE ARE several important differences between lacquer and varnish, some of them advantages and some disadvantages. To begin with, lacquer dries much more quickly. You can turn out a rubbed finish with it in a couple of days—allowing more and better drying time than the can labels ask for. On the other hand, many finishers find lacquer much more difficult to work with. One reason why is the quick drying. Another is a tendency for the lacquer to soften colors of the wood or stains and "bleed" them into subsequent coats. Durability of lacquers is good but some slight deviations from the proper application techniques can cause premature failure of what appears to be a superb finish. The biggest danger lies in pinning too much faith on quick drying. Like varnish, lacquer is better when it has had more than enough drying time between coats.

Lacquer is less used than either shellac or varnish in home workshops—and more used in industrial finishing. The reason is not its appropriateness, for it can be used anywhere a clear finish is called for. However, it takes a little skill to use, and before you attempt a project using it, you should practice a little on a less important job—or on odds and ends of wood similar to that in your project, until you get a feel for the material. Then use it on paneling, furniture, floors (in special formulations intended for floor use), trim—anywhere you want to take advantage of its quick drying.

Lacquer comes in two forms: brushing and spraying. Don't attempt to brush a lacquer intended for spraying; it will dry behind the brush too quickly for anything but a messy job. On the other hand, if you want to spray a brushing lacquer, go ahead. The results are excellent.

The difference between the two is, of course, drying time. Through a careful balance of solvents, the manufacturers can produce a lacquer that dries about the same speed as shellac. An example is Satinlac, which is perhaps the most widely used of brushing lacquers. Another is Fabulon, a finish intended primarily for floors but excellent for use on any wood product. Fabulon is quite a bit harder to use smoothly than Satinlac, and is really best when it is sprayed. However, its build is much greater than Satinlac's, and you arrive at a thick, lustrous finish in a short while.

Excellent brushing lacquer, easy to find at any paint store, is one intended specifically for floors. It handles about as easily as varnish, but dries faster. The key to success with lacquers is longer drying time than labels ask for.

Spraying lacquer in ordinary cans is not often sold in paint stores, but you can order it through professional finishing houses. The ready availability and excellence of Fabulon makes this seem hardly necessary, however.

The well-known spray can, such as Krylon or DuPont Spray enamel clear, is a lacquer, although the film is usually acrylic rather than nitrocellulose. Spray cans of lacquer are inefficient for large-scale work but give good results on small projects. They can be built to a thick finish—but you must never forget that the fast drying time advocated on the label is not meant for multicoat work of the highest quality.

One of the main advantages of a lacquer is that dust becomes much less of a problem. If you clean ahead of your work with a tack rag *used quite dry*, you eliminate dust on the surface. Since the material is dust-free-dry in minutes, there is less chance that dust in the air will settle on it.

Build of floor lacquer is excellent, making it good for rubbed finishes. Material will take a high sheen under buffing compounds. It is excellent for a piano finish (see Chapter 13) or even a guitar finish, as shown here.

107

Excellent but rarely used technique with relatively thin-bodied brushing lacquers is dipping, as shown here with an ebony and ivory cufflink. A good trick is to suspend such small objects on a wire with an L-bent end, then whirl it so that centrifugal force takes off the excess, after a quick drain.

TIPS ON USING LACQUER. Since most people have had much more experience with varnish than with lacquer, the best way to spell out the tricks of lacquer application is to contrast them with varnish.

1. Instead of laying the material on, then smoothing it, you should attempt to flow lacquer on in a good wet coat without too much brush-back. Move fast, using long strokes. Keep a wet edge by working in small areas. Usually a relatively long and narrow area is easiest to handle. To keep the action fast, use a wider brush than you might like for varnish. Never apply lacquer with a tiny brush unless the project itself is tiny. Width rather than fullness is the mark of a good lacquer brush, which needn't have bristles as long as are considered best for varnish.

2. Instead of using lacquer as thick as you can brush it comfortably, as you do with varnish, keep it thin enough to flow out well. This may mean some thinner, even in a material such as Satinlac, which is canned at a brushable consistency as it is. Be sure to buy a good thinner. There are many chemicals which will reduce the consistency of lacquer, but not all of them produce good results. Thin all coats, if necessary.

3. Sanding between coats is not necessary with lacquer *to provide adhesion.* Each coat tends to soften the preceding one minutely, bonding to it. Thus, scarifying for mechanical bond, as with varnish, doesn't help. However, you may want to scuff-sand enough to knock off high spots and the few dust specks you'll be bound to get, even with lacquer. And, if you are working toward a high rubbed finish, similar to the piano finish covered in Chapter 13, you will always *sand to plane* the surface, if application is rough. Do not use water on between-coat sanding with lacquer. Be sure to give lacquer at least four hours to dry before sanding and recoating. Give the final coat overnight, before you do the last stages of rubbing with pumice.

When using lacquer, you may encounter any one of several difficulties. But these can be handled if you know how.

Orange peel. You may get this when you spray if your pressure is not high enough for the viscosity of the lacquer. Instructions that come with spray guns cover the use of viscosity meters and other methods of making sure that lacquer is the right thickness for spraying. Another cause of the pebbled look of orange peel is spraying from too far away, allowing the minute particles of lacquer to harden slightly while they are in the air. You may get this result with spray cans as well as with standard compressor sprayers. Move in slightly, and be sure that you are spraying wet. Bad thinners cause orange peel, too.

Pinholes. Several spraying mistakes cause tiny holes in the surface, caused by vapor-pressure of solvents beneath a surface film which forms quickly. In other words, too much surface drying for the character of the coat. With big spraying equipment, the reason is most often improper thinners. With spray cans, the problem usually comes from spraying too wet. With either type of spraying, you can get pinholes by spraying too soon with the second coat, so that the previous coat is still emitting vapors. These cause, first, bubbles, then pinholes when the bubbles burst.

Sags. Spraying too heavily causes sags and runs, just as brushing too heavily does. However, spraying with too little drying time can be worse. May you be spared the vision of an entire finish softening and sliding off the wood because you sprayed too many coats too soon—and they all softened and sagged at once.

Bleeding. Rosewood, particularly, but also mahogany and some other woods contain pigments which are soluble in the solvents of lacquer. Also, some stains are softened and brought into solution by these solvents. To complicate things further, certain fillers and sealers are softened by lacquer. These are the reasons why you should always use a "lacquer compatible" system throughout. Commercial finishing involves lacquer-base fillers and sealers. You cannot buy these materials at typical paint stores, however. So, the best practices involving easily procurable materials are as follows:

1. Use water or NGR stains. If you find it necessary to use a pigmented wiping stain, give it at least forty-eight hours to dry. Otherwise, the lacquer thinners may act somewhat in the manner of paint and varnish removers.

2. Give any standard paste filler forty-eight hours to set up hard, for the same reason.

3. A safeguard is the use of thinned shellac as a sealer after stains or filler or both.

4. Use a lacquer-type sander-sealer-filler unless the wood is of such open grain that paste fillers are required.

5. Let everything dry thoroughly, in advance of lacquering, especially any oil-resin or petroleum derivative materials.

6. When you plan to use lacquer over rosewood or dark mahogany, use a thin shellac washcoat as a sealer and *do not sand it*. The shellac will seal the pigment, keeping it from bleeding, and you might cut through it if you sand, thus breaking the seal.

Lacquer rubs and buffs to a higher polish than you can get with varnish, which is one reason why it is so often used for small, elegant projects which look best with a high finish. The good rubbing quality also accounts—along with speedy drying—for the popularity of lacquer on most commercial furniture. If you will learn to use it, you'll find lacquer a worthy addition to your finishing techniques for any situation where the extra durability of varnish is not required.

Whenever you are eager to retain the maximum natural wood color— that is, without any of the darkening oil finishes produce—don't forget that lacquer itself is the whitest and the least darkening of all. On top of that is the magical effect of lacquer lighteners, which leave the wood almost identical in color to its raw hue.

CHAPTER FIFTEEN

HOW TO GET A FINE FINISH WITH ENAMEL

FOR MANY people enameling is easy. They might shy away from varnish, but anybody can paint! The truth of the matter is that a bad enamel job looks worse than a bad varnish job, because you don't have the natural good looks of wood to get you off the hook. With enamel, you're putting on a color and if it is uneven, unlevel, saggy, fat-edged, and dustmarked, it looks just plain bad. (Unless you weren't trying, in the first place, and the project barely deserved painting in the first place.)

On the other hand, a fine enamel finish can look just as beautiful as a varnish finish, and many of the highly vaunted finishes of the past were in colors. There is a whole period of furniture and cabinetry which is only proper in enamel—the early American primitive. Most Shaker pieces are better painted than clear. Most old side chairs, carefully stripped and varnished by proud 20th-century owners should have been painted. Some of the finest 18th-century French furniture is finished with enamel. When Chippendale came back from a visit to China, so the history of furniture styles says, he switched much of his styling to "Chinese Chippendale," of which a great many pieces were enameled, then decorated with elaborate chinoiserie brushwork. The lines of the best furniture manufacturers today always contain a few well-styled numbers in enamel.

WHAT IS ENAMEL? Most of us think of enamel as the colored material paint dealers carry in dozens of colors, in cans from half-pints or smaller up to gallons. That is the most common, and perhaps the most useful kind. A good quality enamel of this sort is actually nothing more than a good varnish with pigments added in sufficient quantity to give the finish opacity and color. And, like varnishes, enamels are now being produced in water-thinnable formulas. (See page 92.) This makes possible all the qualities of oil-based enamels with the advantages of latex-type enamels.

The good alkyd enamels, as this might suggest, are handled exactly like varnish, when you're shooting for a good colored finish.

In the paint industry, however, the general category called "enamel" also includes colored lacquers. Again—the material is nothing more than a

Fine finishes, built up coat by coat with enamel, can be the entire equal in color of those done meticulously with varnish. This little jewel box, as pretty as any Chinese lacquer, is made of ordinary fir plywood. (See Chapter 17.)

lacquer formulated with pigments to give it color and hiding power. For most home-workshop finishers, colored lacquer means spray cans of lacquer, such as Krylon, DuPont Spray Paint, and others. Now and then, you'll find a few cans of colored lacquer on a dealer's shelves; it is nearly always spraying lacquer, as is most colored lacquer.

There is, however, an excellent source for high-quality colored lacquers, if you have spraying equipment and want to use them: your auto supply dealer. You can buy lacquer the color of any automobile manufactured in the past ten years or so—just to give you an idea of the color range possible. (As a matter of fact, you can match those colors in oil-resin automobile enamel, which is usually of a quality unmatched in ordinary enamels.) As with oleo-resin enamels, a good colored lacquer is handled exactly like clear lacquer.

Colored brushing lacquer is difficult to find. You can thin spraying lacquers with materials which are slow-drying—relatively—and retard drying time enough so that brushing is possible. Such thinners include amyl ace-

Handle enamel as you do varnish or lacquer, building up a flat finish with wet-sandpaper rubbing between coats.

tate, butyl acetate, and others. However, a colored lacquer thinned enough for easy brushing usually *carries so little color* that you need coat after coat after coat to build both surface and color. For that reason, you should consider lacquer in colors as a spray-only material. Use the spray cans, which give excellent results, or use a regular spray gun.

Enamels come in glossy and semiglossy—and in a variety called "flat enamel" which fits better in the category of paint and which is intended more for painting walls than for finishing wood. The difference between glossy and semiglossy in enamels is basically the same as in varnishes. Pigments break up the surface of the semiglosses so that they have a velvet look.

There are further differences, however. The glossies are normally put up in a *deeper range of colors,* going into dark blues, greens, browns, reds. They are also put up in pastels. The semigloss enamels, on the other hand, are rarely packaged in dark shades. The reason for this is that satinsheen enamels are generally intended to be used as trim colors to match wall paint; there is no real reason why semigloss enamel can't be factory-packaged in the same colors as glossy, other than normal demand. You can,

A good glossy enamel produces a more durable finish than quick-dry spray finishes. How to paint entire piece and let it dry? Trick shown here is to turn three screws (or brads will work) into the bottom. Paint the bottom, then set it on the screws while you paint the rest. When all the coats are on, remove the screws. This box finally got a glued-on felt bottom which covered the holes.

however, obtain semiglossy enamel any color you want in the *custom-color systems* which nearly every paint store has. In very dark shades, semigloss enamels run up in cost, due to the amount of pigment it takes to make them. At the same time, the relatively high proportion of pigments to vehicle makes dark-colored semiglossies less durable—less resistant to abrasion, water, and other damage. Many times, a dark-hued semiglossy is slow to dry, and may take a week or more to harden thoroughly.

For these reasons, semiglossy enamel in dark colors should be your choice only for projects too large to rub, or for parts of a project which are not worth rubbing. To cite an example, you might use a semigloss enamel on the sides and drawer fronts of a chest, but go to the bother of a rubbed enamel finish on the top.

HOW TO DO A RUBBED ENAMEL FINISH. Using a good glossy enamel or a good spraying lacquer or a spray can of lacquer, you can build up a finish that in color is the exact counterpart of a rubbed finish in clear varnish or lacquer. You use exactly the techniques covered under the piano finish sections of Chapter 13, or the rubbed lacquer schedules from Chapter 14.

When you are shooting for a top-grade finish with enamel, your brush-work must be as careful as it is with varnish. Using the same tactics, follow these steps:

First, lay on the enamel in stripes the width of your brush over an area that is comfortable to work with. Have these stripes heavy—twice as thick as you want the final coating—and have them spaced about a brush width apart.

Second, brush out these stripes, crosswise, smoothing the enamel over the entire area.

Third, tip off the enamel with gentle, even strokes lengthwise, letting only the tips of the bristles touch. Hold the brush quite vertical for tipping off, and you'll produce fewer bubbles. When one section is finished, move to the next, and blend them together as you brush out and tip off.

There is a reversal of procedures which you might want to use, however.

As discussed in Chapter 13, you may want to build a short-cut non-gloss finish by putting on two or three coats of *glossy* varnish, then top-coat this build with a semigloss material. The result is a fairly good substitute for a rubbed finish. A short-cut method of building a rubbed enamel is to build level with the extra-bodied semigloss. Rub between coats, until the surface is absolutely plane and smooth. Then put on a coat of glossy enamel for the top coat. Since you do not have to carry the color with this coat, and since the surface beneath it is already quite plane, you can hit the glossy coat with a sanding block and 6/0 paper, then switch to pumice and oil to complete the rubbing. Few finishes of any kind are more beautiful than a well-rubbed build of enamel which is finally coated with a good buffing of paste wax.

Another difference between enamels and varnishes or lacquers is that you do not normally use fillers under them. Partly this is because the pigments in the enamel tend to fill, but of equal importance is the fact that you do not often put colored finishes over the walnuts, mahoganies, teaks, and other woods which are so porous as to require filling for a good, level clear finish. This does not excuse you from a thorough sanding and cleaning of the surface.

ENAMEL UNDERCOATERS. Most enamels specify the use of an enamel undercoater as the first coat. The undercoater has several advantages. It is cheaper than enamel and therefore gives you a first coat at less cost. It's special formulation provides an ideal surface for enameling, and in many cases makes it possible to complete a colored finish in two coats, since you can put more enamel on over an undercoater without the danger of sags and runs.

However, for small jobs, the undercoater is more nuisance than its advantages warrant, and actually represents a cost increase; it's a can you might not have needed. Moreover, it is usually white. This is all right if you are enameling white. But when your enamel is in a color, you must tint the undercoater or you'll never carry a very deep color enamel in one coat. Tinting is usually not feasible deeper than about medium tone. Some workers give the undercoater a little start in the right direction by pouring some of the enamel in it.

As you can see, there are many problems created by undercoaters. On the other hand, enamel itself has tremendous adhesion to a clean surface. If you thin a bit of it and use it as a primer—just as you thin varnish for the first coat—you get perfect results. Save the enamel undercoater for jobs as big as a new houseful of trim and woodwork.

CATALYTIC COATINGS IN COLOR. If you have a problem calling for absolute maximums in protection (such as around sinks and similar situations) you can turn to a type of enamel which is called "catalytic coatings" in the paint industry. These coatings are true epoxies or urethanes which are composed of a resin and a hardener. The two materials are mixed immediately before use. They set and harden chemically rather than by evaporation of thinners only, and when they have cured, their resistance to acids, alkalis, water, and abrasion is unbeatable.

When you buy the two containers, you need not mix them all at once. What you do mix, however, is useless later on, unless you store it in the freezer compartment of the refrigerator. In such low temperatures, the combined resin and hardener stay liquid for a while—but it is best to mix only what you need for a job.

CHAPTER SIXTEEN

THE FINE AND SIMPLE ART OF GLAZING

GLAZING is, by loose definition, the application of a coat of thin paint over a base coat, in such a manner that the base color shows through the top coat. Since one of the most common objectives with glazing is to give a piece of furniture the look of age, the process is often called "antiquing." The base coat may be a color, or it may be a wood tone. The glaze tones and shades the base coat. Quite commonly the base color and glaze, along with the required sandpaper, brushes, etc., are sold in color-coordinated kits. At least one kit—that of Sears, Roebuck and Co.—utilizes water-thinned emulsion materials. Some paint stores sell glazes separately, and since any semigloss enamel works as the base coat, it is simple to put your own color combinations together. Most glazes you buy are in burnt siennas and umbers, and deep brownish ochres. But, a glazed finish doesn't have to stick with the colors intended to make the piece look old. Using pigments in a thinned oil-resin material such as Val-Oil or clear Rez or Firzite, you can make your glazes any color. Some quite stylish finishes result from such combinations as bright red glazed with black or an intense blue glazed with white.

Glazing or antiquing is one of the most popular forms of finishing, since it is so easy to do. You don't need to be any kind of an expert with the brush; in fact, if you are a little inept, it sometimes helps.

APPLYING THE BASE COAT OF ENAMEL. Although a glaze may be put over any appropriate finish that is in good condition, the normal beginning is a coat of semigloss enamel. As long as it covers well, it is not important for this to be a particularly elegant job of enameling; sometimes a few brush marks help give the glazing character. If you are working on a fairly nice piece of furniture, however, you may want to do a careful two-coat job, sanding between coats. It is not a good idea to sand the second coat, however, since the glaze takes best over the natural sheen of the enamel. Sometimes you want the glaze to fill wood pores and accentuate them; when this is the case, thin the enamel slightly, so that it will not "bridge" the pores. You can apply the base coat by spray gun. In fact, if you make up your own color scheme and can find the color you're after in semigloss spray can, it makes a good base coat for small projects.

BRUSHING ON THE GLAZING LIQUID. Applying the glaze is a slap-dash operation. *Be sure to cover completely,* because if you don't, the uncovered spots conspicuously will not tone the same as the rest of the area. Other than that, smoothness of coverage is not critical. Brush the glaze on, swab it on with a rag, blow it on with a spray gun.

Until you get the hang of it, you'll be better off to work on a small area in an inconspicuous spot, so you can learn how fast the glaze dries. The time of drying you allow before you wipe affects the appearance, the ease of wiping, and the degree of control you have over the toning. A little practice helps. (See the color illustrations on pages 127 and 128.)

WIPING THE GLAZING LIQUID. This is the point at which you become particular. The way you wipe the glaze determines the final result. It controls the degree of "antiquity" and it controls the tone of the final color. A garish orange, for example, becomes a tasteful pomegranate under a glazing of deep umber. A green that makes you ill turns into an ochre-olive. The *more* glaze you leave on, the greater is the degree of down-toning. The more you wipe, the less. However, it is virtually impossible to remove all the glaze, and if you do take off too much, merely dab on some more and wipe again.

Although you should wipe the glaze any way that seems attractive to you, there are a few tricks proven effective by experience:

1. Wipe hardest on high spots—corners, edges, knobs, moldings, and so forth. In a sense, you wipe the glaze as though *wear and use* caused the erosion of the glazing. Any place where normal use might have worn the finish thin, wipe hard.

2. Conversely, wipe least—even not much at all—in depressions—carvings, low spots in turnings, grooves in moldings.

3. Wipe the center of a large area, such as a table top, drawer front, chest end, etc. Try to blend the glaze from quite light at the middle to fairly heavy toward the edges, but light once more at the very edges.

4. Spend a little time on intricate carvings and turnings; you can add to their decorative quality by spot-wiping highlights with cloth over the end of your finger, and by using cotton on a stick to wipe crevices.

5. When the object is accentuation of the grain (as in the case of the green table in the picture) or of attractive old finish deterioration, wipe *across* the grain or cracks with a *thin* pad of rag, so that it doesn't dig into the fine cracks, but leaves the glaze there.

6. Sometimes a piece calls for special wiping techniques. For example, turkish toweling, used carefully in long, parallel strokes, leaves a subtle striping of glaze. In other cases you can use stippling tricks—a wadded

newspaper, or the stiff ends of a stippling brush. Or, you can use a stiff brush with a stroking movement to create a grain pattern.

7. Since you may take more glaze off than you want to, *stop* when you think you've gone far enough. You can even let it sit overnight if it is nearly finished. In the morning, take a new and fresh look. If you want to remove more glaze here and there, you can do it quickly with very fine sandpaper or fine steel wool. It is not even difficult to take it all off with turpentine, if you do it soon enough after you decide you want to start over; the glaze is not a particularly rugged finish in itself, and can be wiped off with rags and turpentine or paint thinner when it is fairly fresh.

8. If you topcoat the antiquing job with a durable varnish, wait two or three days for the glaze to harden, and then test a spot to see if the varnish softens the glaze. In some cases, it may. You'll find the super-clear urethane varnishes affect the color of the glaze very little, and provide a surface that will withstand a lot of rough use. Pick a semigloss or a flat urethane unless there is some reason why the final finish should be glossy.

Although nine times out of ten you may want to do an ordinary wiping job with a wadded rag, there are several techniques for producing special patterns and textures which may suit the piece and your mood:

Brush-tipping. After you have wiped most of the glaze, use a fairly stiff brush in long straight strokes to provide the look of wood grain.

You may wish to apply a wood-grain finish over your painted cabinets or woodwork. If so, there are special glazing kits for this purpose. Wiping the glaze down with steel wool gives you a wood-grain effect.

Stippling. Dab at the partially wiped *glaze* with the ends of the bristles.

Textured. Wad a piece of newspaper or plastic film and pat it over the surface.

Marbelizing. Lay a sheet of Saran wrap in the glaze and crumple it a bit. Then peel it off.

Spatter. Instead of brushing or swabbing the glaze on, flip it off the end of the paint brush by striking it over the edge of your hand, or bending the bristles with a stick. Try a broom, a tooth brush, a vegetable brush, for different effects.

BUILDING YOUR OWN KITS. The following is an example of what you get in an "antiquing kit" at your paint store:

Small can of semigloss enamel	Wad of steel wool
Smaller can of antiquing liquid	Wad of cheesecloth
Small throw-away paint brush	Sandpaper in three grades
Smaller brush for the glaze	Step-by-step booklet

Because you have everything you need in the kit except the durable varnish for possible protective coating, you may want to buy your glazing supplies this way. Or, you may want to buy the actual finishing materials and use sandpaper, steel wool, cloth, and brushes from your own stock. In some paint stores you can buy semigloss enamels specifically for glazing, along with glazes color-coded to go with the enamels. Many workers prefer to put together ingredients using their favorite materials. Buy a good grade of *satin-finish* enamel. Pick up a tube or two of pigments, black, ochre, burnt umber, or whatever colors you'll need, along with a small can of a clear resin sealer. From these, mix the glazing material.

Important: Use colors in *oil* if you can find them. If you must use the universal colorants that are more common, give them an hour or two after mixing, with occasional stirring, for the "universal" liquid to evaporate away. It will not take long, for it is a highly volatile methyl alcohol in most cases. On the other hand, if you do not let it evaporate out, it may be strong enough to soften the base coat.

PREPARING THE SURFACE FOR A GLAZED FINISH. One of the features of glazed finishes which appeals to most people is that there is nothing gained by removing an old finish, as long as it is in sound condition. Blemishes, cracks, crazing, and other character marks typical of old varnishes and enamels often pick up the glaze and because of the emphasis becomes pluses instead of minuses in the finish. However, if there are ugly scratches that are finish-deep or other blemishes which won't stand being highlighted, get rid of them. Either remove the finish or spot-sand until the blemish disappears. Remember, you are not concerned with any discoloration; it will disappear anyway.

In some cases, you may want to make deliberate blemishes. If there is a lacking of character marks, a few strokes with medium or coarse sandpaper produces a texture for the glaze to catch in.

Many times, the drawer pulls and other hardware on a piece of old furniture are all out of key with the look of a glazed finish. Anticipate this, as was done with the oak chest turned blue in the photographs. Remove the old knobs, plug up the holes with wood putty or other filler and sand them into invisibility. Then, when the new finish is hard, drill new holes and apply the new and more appropriate hardware.

One thing is sure: any old finish is *dirty*. Wash it with a detergent, or use a product like Wilbond or Liquid Sandpaper, to be sure the old surface is ready to take a new finish. If the old finish happens to be high-gloss—which seems unlikely—scuff sanding or "chemical" sanding will produce enough tooth for the new enamel to stick to.

WHEN A GLAZED FINISH IS BEST. Glazing has a special place in the world of furniture finishing. Its most useful function is in bringing junk down out of the attic and putting it back to useful life. Generally, this means furniture which falls short of being stylish—but which gains the look of distinction because of the interesting finish you put on it.

Antiquing as a name for glazing doesn't change that. It does not belong on fine furniture. It doesn't belong on good antiques. It doesn't even belong on good wood, except when the good wood is in a bad piece of furniture which can be made good with a glamorous finish. The pieces which illustrate this chapter—all from secondhand stores—are typical of the kind of furniture that benefit most.

Another classification is the old piece of furniture of good design which may not be made of good wood and is not worth the bother of a clear finish. A great deal of Chinese Chippendale, Hepplewhite, Sheraton, and other attractive styles were manufactured during the period from about 1920 to about 1940 of pretty bad wood. The objective was to meet price, and the cheapening included short-lived lacquer finishes. When you strip the finish off these pieces, it breaks your heart to find that the wood is bad—mismatches of narrow boards glued up, cheap veneers with gum or poplar at the edges. A glazed finish is the best treatment for such pieces—and with a little practice, your eye will come to recognize them before you go to the bother of stripping them.

The glazed finish does not blaspheme the names of those great 18th-century designers. A Chinese Chippendale credenza done in red with a black and umber glaze wouldn't cause a stir in Chippendale's grave. Tasteful colors on Sheraton and Hepplewhite, on Adam, on Beidermeier are correct. Much good 18th-century French furniture was colored and glazed —usually in ivory with umber. Much bad furniture of American's great golden oak age looks mighty close to French, with a good glazing job. Fool almost anybody with a little fine-brush decoration, gold edging, and stripping.

CHAPTER SEVENTEEN

HOW TO FINISH FIR PLYWOOD

ONE of the most popular materials for home workshops is fir plywood. It is easy to work with, putting projects within the reach of everybody which might be too difficult or too expensive working with some other woods. Well handled, fir plywood can be made to look very attractive. But—there are a few tricks in finishing this wood.

The special problems you encounter with fir plywood arise from the way the plywood is rotary sliced—around and around the tree. This produces the characteristic pattern of light softwood and dark, dense wood. These two kinds of wood accept finishing material in drastically different ways. Also, since the sliced wood comes off the tree in a huge, curled, cross-grain shaving which is flattened in lamination, there are unavoidable stresses which reveal themselves as checking. Take a look at any piece of plywood that has been lying around for a while and notice the fine, parallel, grainwise cracks. Control of these is the second problem in finishing firply well.

GRADES OF FIRPLY. The first step toward a good finish is selecting the right plywood for the job. At the mill it is graded A, B, C, D. The A grade is flawless. B may have some boat patches. C may have some knots. D may have some knotholes. Designations such as A-C or B-D mean that one side is faced with one grade and the other—the back—with a lesser grade.

There is no sense in buying a better piece of firply than you need for the job. A functional project is just as solid built of B or C as it is of A. However, the better the grade, the easier it is to finish. B grade's boat patches and frequently narrower strips of veneer give you problems at finishing time which may make you wish you'd spent the money for A grade. You can not hide patches under any clear or stained finish, although enamels cover them. If your project is to be stained and it is supposed to be nice, you need A grade.

FIRPLY SANDING TRICKS. When you buy firply it is plane, but it may not be as smooth as you'd like it. At the mill they drumsand it down to the equivalent of about 3/0. If you want to make it smoother, follow the following instructions:

The first step in finishing plywood is to examine the surface for blemishes which must be filled. Grade A will have none, lower grades will show boat patches. No matter how carefully they are fitted, a smooth finish over them demands filling with a fine grain filler such as vinyl spackle or a water-mix putty.

1. *Never* sand firply without a sanding block *big enough to bridge across the soft, light-colored areas.* They are so soft that they wear hollow if you use a small block—or paper held in your fingers.

2. If possible—and this means nine times out of ten—apply the first coat of finish before you sand at all, to act as a sealer. Many workers prefer to use a regular sanding sealer on firply. The reason: sealer or the right finish coat hardens the softwood areas and helps keep them from dishing. Once hardened, the fibers can be made smoother too. Otherwise, there is little point in going finer than 3/0; the dark, dense areas may get smoother, but not the light, soft areas.

HOW TO CONTROL CHECKING. Firply won't check if it is properly sealed against moisture. Do not expect to get sufficient priming and sealing

Edges of plywood always show voids or torn fibers. Fill them with vinyl or water-mix putty. Work the filler into depressions with your finger, making sure to overfill. Then clean off the excess with sandpaper.

from ordinary primers or from the first coat of a self-priming material. The best results come from soaking firply with Firzite, Pentaprime, Rez, DuPont Penetrating Wood Finish or other similar materials which soak into the wood and fill the cells and inter-fiber spaces with resins. Apply these materials only to the *bare wood*, never over sanding sealers.

Since moisture readily enters the edges of plywood, you must seal edges which will butt against other surfaces *before* you assemble the joints. If the joints are glued with a material equal to Weldwood Plastic Resin Glue, you get a good seal from the adhesive if you apply plenty of it to the edge.

On projects destined for sheltered life, you have less checking to worry about. Nevertheless, one of the above penetrating sealers is good insurance, even for a dresser top jewelbox with a rubbed enamel finish.

HOW TO TAME WILD GRAIN. The colored photographs illustrate the garish results you get when you stain firply in the raw. By soaking up

This photo accentuates the checking which is characteristic of all smooth fir plywood unless it is carefully sealed against the moisture which causes it. There are special sealer-primers for ply-wood, and unless they are used you may end up with rough surface like this. Note how the boat patch, also, is emphasized when sealer is not used.

quantities of stain, the softwood areas go dark, while the denser wood stays about the same. The contrast is too great for good looks.

The same standard firply sealers help you kill wild grain by reducing the capacity of the softwood to absorb stain. Thinned shellac works in the same way, but without the protection against checking. Some workers thin White Firzite with turpentine and use it as a brush-on-wipe-off seal. The soft areas are not only sealed, but they retain enough white pigment to hold the stain lighter in the places where it would normally be darkest.

PAINT AND ENAMEL FOR FIRPLY. The uneven character of firply's grain presents less trouble under opaque finishes than under clear—but the trouble is still there. If you want to build a smooth enamel on fir plywood, start with a penetrating seal, to suppress checking and to level off absorption. Be particularly careful not to let sandpaper dip into the softwood areas, because any unlevelness is all the more conspicuous with the uniform color of enamel.

Handle the enamel carefully to avoid heavy spots, working up to level with two or more coats, each scuff-sanded thoroughly. Avoid dust, particularly on the final coat. If your project warrants it, a third or fourth coat of enamel wet-sanded with 400 grit, then waxed, is the equivalent of "Chinese lacquer." (Spray-can lacquers can be built the same way, if you temper the "quick dry" fast talk and give each sprayed coat a minimum of three hours in good drying conditions.

Paint—in the sense of a utilitarian finish—calls for slight modification. Outdoors, where much fir plywood is used in residential construction, experienced workers often apply the seal and the first coat. Then, in six months or so, when checking is more or less final, they put on the remaining coat—or two, if the schedule calls for it. This technique works just as well with semigloss enamels indoors.

TEXTURED PLYWOODS. Firply is made in dozens of textures such as the Surfwood and Weldtex of U.S. Plywood and the industry-wide "Texture 111." The swirled or striped surface of these textures obscures both the wild grain and the tendency to check. In addition, it provides opportunity for some interesting novelty finishes.

The trick used for most of these finishes is a base color, over which you apply a second color in some manner which takes advantage of the texture. These techniques include:

1. Spraying at a flat angle, so that color is deposited only on one side of the texture.

2. Brush-on-wipe-off, leaving color in the depressions and more or less removing it off high spots.

Since water can get into plywood through the edges and cause checking, be sure to seal all edges which butt against other surfaces. This must be done before assembly, of course, and an easy way to speed the job is to stack the pieces and prime them all at once.

3. Applying the second color with a hard roller which bridges the low spots and colors the high.

Additional interesting effects are possible with single colors, especially on Surfwood, which is a sandblasted surface. Much of the softwood is removed, and what remains has a pebbled surface. When you brush on a semiglossy enamel, it holds its sheen on the hardwood but looks quite flat over the pebbled surfaces. As with several others, this novelty finish on textured firply seems to shift and change as you move past it and angles of light change.

FINISHES FOR FIRPLY. As the color swatches on pages 129-131 suggest, there is an almost endless variety in the finishes which are suited to fir plywood—or which are possible only on fir plywood. Here are the processes used in producing each of the finishes shown:

Pigmented wiping stain. This type of stain shows a great deal of grain contrast with the odd effect of darkening the light grain so much it becomes the dark grain. For this sample, a walnut stain was used in the ordinary manner—brushed on, allowed to penetrate for a few minutes, then wiped off.

Emulsion wood stain. This material handles exactly like an oil-base wiping stain, except that its thinner and brush cleaner is water. As you can see, it accentuates the grain to a slightly lesser degree than the oil-base material, although the contrast increases slightly when you put a topcoating of varnish over it. This material, called Deft, is good under lacquer topcoatings, showing small tendency to lift or bleed.

Presealed non-grain-raising stain. This sample of fir plywood was sealed with clear penetrating resin, then colored with non-grain-raising stain. To control the degree of penetration, wipe the NGR, although it is common to let the NGR dry when you want darker tones. As you can see, the sealer toned down the contrast between light and dark wood.

Varnish stain. When you apply a varnish stain in the normal manner, the softwoods go quite dark, while the harder areas stay light. This is a fairly common finish for plywood in noncritical areas where some protection is important but a great deal of beauty is not.

Wiped varnish stain. Good control of wild grain results from wiping a varnish stain as though it were a wiping stain.

Seal under water stain. Penetrating seal on fir plywood, followed by a water stain subdues the wild grain almost completely.

Water stain. Water stain over bare wood gives sharp contrast. In the lighter shades, it is less contrasty than in darks, since the hardwood grain soaks up relatively little color, regardless of depth of stain.

White wiping stain. This is regular white Firzite, a commercial white sealer for fir plywood, used as a wiping stain.

Ebony wiping stain. Darkest available pigmented wiping stain is ebony, which gives fir plywood an interesting black-and-gold look. Give the stain plenty of time to sink in.

Stained textured firply. This finish takes advantage of the textures of sandblasted fir plywood. To do this finish, first seal with thinned shellac. Then add a little turpentine to a semigloss enamel and brush-on-wipe-off as you would a pigmented wiping stain. Green and gold is an interesting effect for paneling.

Semigloss enamel. Over sealed textured plywood, a semigloss enamel produces a satiny look over the high, hard areas, and a pebbled, dull look in the soft area. This effect is best with fairly dark colors.

White stained texture. Another good paneling treatment, particularly for dimly lighted areas such as basements, is white stain over a shellac seal on textured plywood. This treatment lightens the overall color without any increase in grain contrast.

Angle-sprayed textured plywood. To produce this effect, hold the spray gun close to the surface, but at a flat angle. The spray hits one side of the texture and colors it, while the opposite side stays light.

Enamel striated plywood. As this sample shows, striated plywood shows none of the wood-grain pattern normally associated with fir plywood, even when it has only a coat of seal and one coat of enamel or paint. Use two coats on porches and other outdoor locations, or whenever you find it necessary to get good coverage.

To achieve a handsome glazed finish, three basic steps are involved: (1) applying the base color, (2) applying the glaze, and (3) wiping off most of the glaze as shown here. You can obtain the base color and glaze from your paint store. Or you can use one of the many kits available today, such as the one pictured at right. When the glaze is wiped away properly, the result should be an antique appearance simulating normal wear after long use. Wipe most of the glaze from flat areas and high spots of moldings or ornaments, leaving a considerable amount in the depressions.

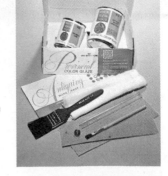

127

The first step toward achieving a glazed finish is to apply a satin-finish enamel or a base coat from a glazing kit.

After the base coat is dry, the glazing liquid is then applied. Cover the surface well, but don't worry about irregularity of brush strokes. Begin wiping with cheese cloth when the glaze starts to dull over.

Leave a certain amount of glaze in the indented areas of ornamentation to emphasize antique appearance. Use a toothpick wrapped with cotton or small medical swab to touch up these areas.

A spray-can product is particularly useful as a base coat on picture frames and heavily ornamented small furniture. Make sure the base coat is thoroughly dry before applying the glaze.

Angle-Sprayed Textured Plywood

Enamel Striated Plywood

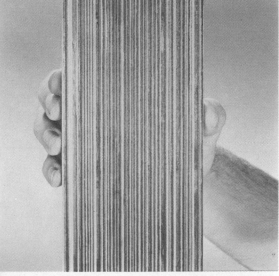

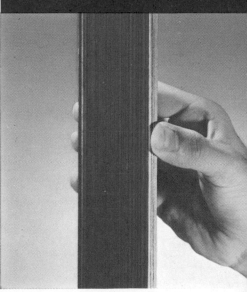

(Note: Color photos on this page and the next two pages illustrate principles explained in Chapter 17.)

Striated plywood, when sprayed at an angle, changes color and character when your view of it changes. Straight-on view, at upper left, shows equal combination of color and wood finish; view from sprayed angle, upper right, shows nearly solid color; opposite angle reveals mostly wood color. Two colors, sprayed from opposite angles, also give striking effect.

Pigmented Wiping Stain

Emulsion Wood Stain

Presealed Non-Grain-Raising Stain

Wiped Varnish Stain

Varnish Stain

Seal Under Water Stain

Water Stain

White Wiping Stain

Ebony Wiping Stain

Semigloss Enamel

Stained Textured Firply

White Stained Texture

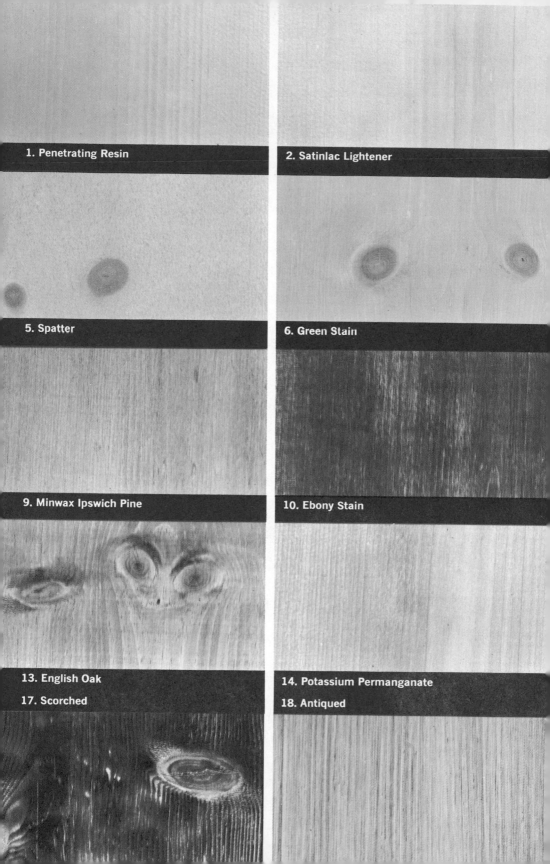

1. Penetrating Resin

2. Satinlac Lightener

5. Spatter

6. Green Stain

9. Minwax Ipswich Pine

10. Ebony Stain

13. English Oak

14. Potassium Permanganate

17. Scorched

18. Antiqued

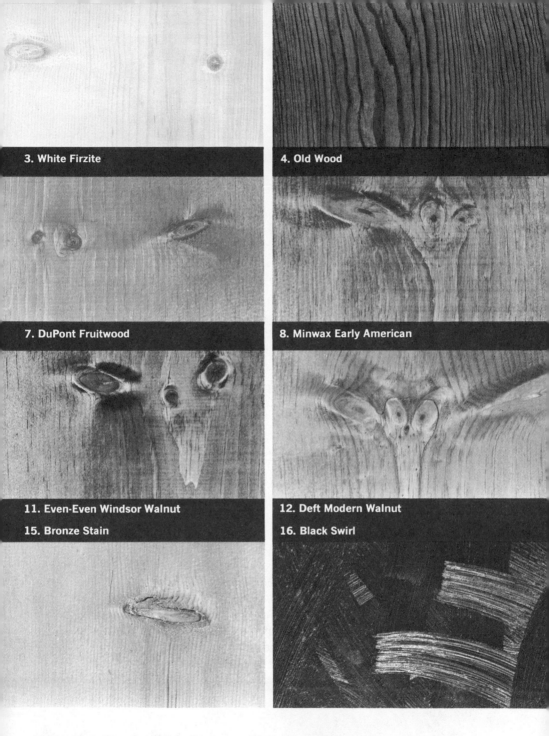

3. White Firzite

4. Old Wood

7. DuPont Fruitwood

8. Minwax Early American

11. Even-Even Windsor Walnut

15. Bronze Stain

12. Deft Modern Walnut

16. Black Swirl

What You Can Do With Common Pine (See Chapter 18 for details.)

There are many finishes available today to help you achieve just about any result you wish with common pine. Here are 18 examples in full color.

CHAPTER EIGHTEEN

HOW TO FINISH PINE

SOME of the problems you run into finishing fir plywood are there when you go to work on pine. Wildness of grain must sometimes be subdued—and you use the same type of sealer to do the job. Sometimes pine shows a tendency to check, particularly around knots. If it is carefully dried and not too wet when sawn and planed, this checking is insignificant. Sometimes it gets completely out of hand—and you have a piece of scrap. Other times you ignore the checking and let it be a part of the character of the wood.

One or two problems with pine are different from fir. Now and then you have sapstreaks or pitchy knots which must be sealed. Sapstreaks are areas in the wood where, during the growth of the tree, pockets of pitch accumulated, saturating the wood. If you put finish over these spots, the solvents in the finish soften the pitch—and the finish never dries.

HANDLING SAPSTREAKS. The best thing to do with sapstreaks is to use a solvent (paint thinner, turpentine, etc.) to soften and wipe away as much pitch as possible. Flood the area with solvent. Let it soften a while, then wipe it dry. Repeat this two or three times if the streak is particularly bad. When it dries hard, brush on a coat of thinned shellac—at about 1-pound cut. Feather this application out at the edges of the sapstreak and avoid going beyond it more than necessary unless you are shellac-sealing the entire surface for other reasons. Since the alcohol solvent in shellac doesn't soften the pitch, it remains sealed under the shellac when you put on other coatings. Remember, however, that some of the urethane varnishes and some of the emulsion finishes do not adhere to shellac. Read the labels.

HANDLING KNOTS. The knots in pine are invariably pitchy, and the area around them usually is too. Treat them the same as you do sapstreaks. However, if you do plan to stain the wood, you may not want shellac to interfere with the action of the stain, which is often most attractive around knots. If that is the case, do either of two things:

1. Carefully shellac-seal the knot only. This will take care of the pitchiest areas. The rest, most often, is only moderately filled with pitch and can be counted on to stay dry under a couple of topcoatings.

2. Stain the work before you seal, if there is not too much pitch. If there is quite a heavy deposit, wipe it two or three times with a paint thinner or turpentine or a multisolvent such as Cleanwoode.

BEST FINISHES FOR COMMON PINE. People who think that pine is for shelves and boxes and benches enjoy woodworking more once they make the discovery that pine treated with respect is the most useful and versatile material in the woodworking world. Some of today's finest furniture is made of pine. Many antiques in museums are pine. And the warmest, most friendly rooms in your neighborhood are probably paneled in pine. (See samples on pages 132-133.)

Most important, many of the *best tricks* in the wood finisher's bag will work on pine—and only on pine.

Just what can you do with pine? To find out, study the color photos of actual pine finishes in this chapter. To make those samples, ordinary No. 2 pine shelving was used, right off the lumber dealer's pile. The pine was sawed into eighteen pieces about a foot long, carefully selected for uniformity in the raw, so that you can judge fairly truthfully the effect produced by each finish. (One piece is not identical, for obvious reasons—the aged pine.) Here is the data for each of the samples, along with comments about the difference in appearance which you should particularly note:

1. **Penetrating resin** intensifies the colors in pine, emphasizes grain. The wood looks as if it has no finish at all except for a slight darkening, but protection is excellent. In time, the wood darkens naturally. (Clear Rez, Firzite, DuPont Penetrating Finish, standard penetrating floor finishes can be used.) Brush it on. Keep the wood wet with additional finish for about half an hour, then wipe it all off. *Do not allow these finishes to dry on the surface.* If the wood is absorbent, repeat application after twenty-four hours. (See Chapter 7.)

2. **Satinlac Lightener** is a special lacquer formula that coats the wood without giving it a "wet" or oiled look. The color of the wood barely changes, although it darkens somewhat with time. Some porous wood may need two coats to seal it. The lightener is intended for use under lacquers, which are quite white, and do not yellow the finish. Don't put varnish over the whitener unless you don't mind the slight yellowing most varnishes give. Pryme is another effective wood lightener.

3. **White Firzite** is, in effect, a white stain, as you have seen in previous chapters. You brush it on, wait twenty to thirty minutes for it to penetrate, then wipe off the surplus. Softer grain areas absorb most of the pigment so the grain is intensified. It is often used without further finish on paneling. For hard-wear areas, it should be topcoated with clearest, palest varnish, such as a satin-finish polyurethane.

4. **Old wood,** which is naturally aged pine, looks gray and lifeless, but regains beautiful deep reds and browns under boiled linseed oil. It takes a few months in rain and sun to produce the aging, making this the slowest finish there is. But no other technique produces this color, although penetrating resin finishes are closely similar. Handle the oil just as you would a penetrating finish: brush or swab it on, keep it wet for half an hour or so—then wipe all the oil off the surface. Or you can use the standard linseed oil finish covered on page 105. If you want extra protection use the oil to produce the colors in the aged wood, then brush on a coat or two of varnish after the oil is thoroughly dry.

5. **Spatter finish** is one you often see on contemporary furniture, done with a base coat of woodtone brown, spattered with dark brown or black. As shown here, green is spattered over pine sealed with shellac. Other possibilities are the use of one spatter color over a base coat of another— such as medium blue over wood-tone brown. Spatter finishes give the impression of one color from a distance, and gain extra interest from the way the two colors separate as you examine them. (Some finishers produce good-looking results by spattering two or more colors.) The basic technique is to dip a brush only an eighth of an inch or so in paint, then flip the paint on the wood. Practice on a piece of newspaper until you perfect the dipping-and-flipping technique.

6. **Green stain** is the same as shown for plywood. Most often used colors are greens, blues, and other "cold" hues not possible with regular stains. Some finishers use wall paint for the stain, since it is less sticky to handle than stain made from enamel, and a slightly different effect is produced by the coarser pigments in paint.

7. **Fruitwood stain,** shown here in a DuPont shade, is an example of dozens of woodtone stains made by all major paint firms. Colors of stains on dealers' shelves change from year to year, as fashions in furniture change. This particular stain became popular in the 1950s, when much commercial furniture was finished in a not-red, not-brown wood hue which they called fruitwood. Handle this type of stain like any pigmented wiping stain—which is what it is.

8. **Minwax Early American,** another wiping stain, is one of the best darker brown colors available. On the wood it is a combination of color that dyes the fibers and pigment that wipes into and emphasizes the pores.

9. **Minwax Ipswich Pine** is lighter than Early American, and a companion of Puritan Pine, which is still lighter. The wiping stains with thin resin-formulated vehicles provide enough protection for small-abuse areas such as paneling, although many users topcoat them with a clear finish. On hard-wear areas like table tops, it is best to follow with a two-coat varnish schedule.

10. **Ebony stain** makes pine black. If the wood is flat-grain-sawed, some of the annual rings hold a golden color. Then pine leaps out of the ordinary and becomes a luxury material. Small furniture pieces, a jewelbox,

a wall bracket, free-standing shelves done in black-stained pine take on the look of ebony and onyx. They look best topcoated with a semigloss or flat varnish to remove any shine. (Compare the way ebony stain looks on pine with its effect on plywood.)

11. **Even-Even Windsor Walnut** is great on pine, despite its name. This stain comes in a tube, a toothpaste-like material that you smear on with a cloth or special dauber. As long as you leave no stain film on the surface, uniformity of application is almost automatic. Take pains, however, to cover entirely, leaving no unwet spots, for the stain has no "creep," as oil-based ones do. There is little or no protection in the water-emulsion stains, so they must be topcoated. Shellac, lacquer, varnish are all compatible.

12. **Deft Modern Walnut** shows by its contrast with Windsor how little names mean when you buy color. It is one of the water emulsion wood stains that you apply with a damp rag. These stains are nonflammable. Their colors approach those available with the best watercolor stains used in industrial wood-finishing techniques. Generally, the color seems flat, dull, and dusty, when the stains are dry. The topcoating brings out the true color.

13. **English Oak** is a custom-mixed stain from the Colorizer line. The nearly black color, with golden brown showing through, is excellent for Early American pieces. Well-stocked paint stores now carry stains as well as paints and enamels, so you can mix or intermix any stain color you want.

14. **Potassium permanganate** (poison), which you can buy at photoshops or drug stores, is a chemical oxidant that turns pine a shade of brown you can't get any other way. The permanganate is red-purple in solution in water and is a potent stain. A standard way to use it is to mix up a saturated solution of the crystals in water. You take a relatively small quantity of this stock solution and dilute it with the amount of water you need for the stain job. If it is too light, repeat applications are cumulative. Flood the surface quickly and uniformly, to avoid mottling. Topcoat with shellac, varnish, or lacquer.

15. **Bronze stain** is an interesting color for small pine pieces. It is nothing more than a bronzing solution, which you mix yourself. Stir powdered bronze into a penetrating resin such as clear Firzite or Rez. Keep it in suspension by frequent stirring and brush it on, wipe it off.

16. **Black swirl** utilizes the economy and workability of pine as the base for a built-up finish sometimes called "gesso." To apply the finish as illustrated here, you smear plaster of paris smoothly over the surface, then texture it with a fairly stiff, dry brush. Any sort of texturing—random or deliberate—is possible. When the plaster is hard, apply enamel. Try to cover in one coat, if you want to maintain sharp textures; go to two or three if you want to round off the edges. Some paint stores sell special gesso formulas.

17. **Scorched finish** is another novelty texture good for small work.

Use a propane torch or other flame to scorch the wood just enough to blacken the surface. Because of varying hardness in the grain, the depth of the burning varies. Then a wire brush takes away the charcoal, leaving color and texture ready for varnish or lacquer.

18. **Glazed finish** is excellent on pine, often used to emphasize the pattern and texture of knots. As shown here, the glaze has a striped look produced by long, parallel wiping strokes with a piece of toweling, after most of the glazing liquid was wiped off.

CHAPTER NINETEEN

HOW TO FINISH AND REFINISH WOOD FLOORS

A FLOOR is quite likely the largest area of wood the average person ever tackles. In addition, it represents one of the biggest durability problems of all; people walk all over floor finishes, yet expect them to look decent. These two differences spell out the two techniques that differentiate floor finishing from furniture finishing. First, you use big, fast-working equipment. Second, you use the toughest materials you can find. Aside from that, finishing a floor is like finishing a dining-room table.

MODERN TRENDS IN FLOOR FINISHING. Finishes for floors have undergone a great transformation. In the beginning, homeowners put oil on their floors, and paid the price of sticky, hard-to-clean surfaces. When homes got nicer and flooring began to take its present tongue-and-groove forms, varnish became popular for its glossy, well-kept look, and people began to pay the price of yellowing, darkening, scratching, and frequent refinishing. Over the years the varnishes got better, and the oils got better. The oil became today's magnificent penetrating floor seal. The varnish became today's tremendously abrasion-resistant urethanes.

Meanwhile, fashion lost its love for shiny floors, and a satin look became popular. And, fashion tired of oak and maple floors the color of oak and maple. In came woodhue stains, and a high percentage of today's floors are stained a medium or dark color more like walnut than any other shade. In new construction and in much remodeling, factory-finished flooring is gaining rapidly in popularity, and there is considerable basis for the opinion in the flooring industry that before long all wood flooring will be factory finished.

FINISHES FOR FLOORS. The finishes you put on floors are essentially the same as those you put on any other wood—penetrating resin sealers, shellac, varnish, lacquer. They are formulated specifically for floors and they are sometimes applied in a slightly different manner than are furniture finishes.

Penetrating wood finish. Every paint manufacturer has a penetrating finish that is excellent for floors and possibly intended also for other finishing jobs. In addition, there are firms that manufacturer floor finishes only—specified and formulated for floors only—usable on other wood, but not intended for it. Quite often these specialty finishes can be bought only from flooring contractors—or in some cases may be applied by contractors who will sell the materials only with their services bargained in. Look in the Yellow Pages under *flooring contractors,* if your paint or lumber dealer can't supply you with penetrating floor seals.

Among flooring manufacturers, contractors, and other experts there is almost universal agreement that the penetrating materials are the best finishes for floors. As Chapter 7 explained, the penetrating materials leave their tough resins *in* the wood, not *on* it. These resins harden the wood, making it more wear-resistant than if it were raw. Since they sink into the wood, the finish is actually measurably *deep,* and to wear off the finish you must wear off the wood which contains the finish. Much prefinished flooring has penetrating resin sealer as its finish.

Shellac. More shellac is sold for floors than for any other purpose. It has the advantage of flexibility, economy, and quick drying. When it is properly applied it wears well, except under hard-use situations. One of its best characteristics is "patchability." When a spot becomes worn, you can sand and clean it, then feather in a new shellac finish in a patch you can hardly see. (It is impossible to do this with varnish, although penetrating finishes are easy to patch.) The major disadvantage of shellac is a low resistance to water and other liquids. It must usually be carefully waxed—because much of its durability comes from the wax.

Varnish. Although the first coat of varnish may penetrate into the wood to a degree (particularly if it is properly thinned about 10 percent with turpentine or paint thinner), varnish is fundamentally an on-the-wood finish. When it wears or scratches off, it reveals bared wood. Patching is difficult. Abrasion resistance, however, is very high, particularly among the urethane formulas. Special high-build varnishes are made for floors, under the name of "gym" varnish, usually. These materials tend to yellow more quickly than the urethanes, which build less while providing superior wear resistance. Each coat of varnish takes at least twenty-four hours to dry.

Lacquer. The wearing characteristics of lacquer are about equal to varnish. Its advantage lies in quick dry. You can get two or more coats of lacquer on a floor in one good drying day.

Quick-dry materials. In addition to shellac and lacquer, there are other quick-drying materials sold at paint stores. Most often they are vinyl-toluene formulas. They are not very abrasion-resistant, and some of the manufacturers stipulate that floors are only a secondary use for the materials.

Special varnishes. The urethane varnishes which are labeled "oil-modified" are *air-drying* materials. Their solvents pass off in vapors, leaving a film. Another form of urethane varnish is *moisture-cure*. It doesn't dry, in the ordinary sense. Rather, its film, taking up minute quantities of moisture from the air, cures and hardens. There are, also, catalytic-type varnishes to which a dryer or hardener is added just before use. Paint stores may not carry these materials in stock unless they happen to be regular suppliers for flooring contractors. The reason is that most home-owners wouldn't want to use them. They are, however, entirely within the capabilities of any moderately handy wood finisher, and have many advantages in durability. Ask your paint dealer to order them, if you want their advantages.

WHICH FINISH IS BEST FOR YOU? When you select a finish for a floor, you must do it on the basis of the most important considerations:

From the standpoint of *wear*, the best on-the-surface material is a moisture-cure urethane. This stuff is so tough they use it on warehouse floors, where it takes the beating of wheeled moving equipment. It has a surface about as glossy as varnish, a good color, only slight yellowing. It has a disadvantage, however. To make the material brushable, they usually use xylene. The vapors are toxic. Without plenty of moving air, they could produce unconsciousness. Never use them without ventilation—and that means more than just cracking a window an inch or two. Open two doors, to be safe.

Oil-modified urethanes or "polyurethanes" are only slightly less wear-resistant than their moisture-curing cousins. They come as satin or semi-gloss, and for that reason are a prime choice for fashionable floors.

The best finish *in the wood* from the wear standpoint is the special floor sealer mentioned above. When you use one of these finishes prop-erly, there is just about no reason why the floors shouldn't look good for the life of the house, with ordinary care.

Best from the *beauty* standpoint depends quite a bit on the eye of the beholder. The plain, untouched look of in-the-wood finishes is considered loveliest by most people, followed by the satin look of the nonglossy ure-thanes. Many floor experts combine penetrating seal with a nongloss var-nish, getting advantages of both. The in-the-wood finish proves an excel-lent base for the satin-finish urethane. Much of the patchability of the in-the-wood is gained by this method, too, since the wood *under* the varnish carries the varnish color deeper than just the surface.

Best as to *ease of application* is, again, the penetrating finish. You merely brush it on, let it sink in a while, then wipe it off. Second easiest is shellac, applied as several coats. Lacquer is harder than shellac, but less difficult than varnish. Most difficult of all are the moisture-curing or catalytic varnishes. The actual difficulty of use is no greater with these ma-terials than with ordinary varnish; the problems they offer are the need for ventilation with its attendant perils, and the chance that the materials may harden before a moderately slow worker can handle them.

Best as to *maintenance* can be decided only by dividing maintenance into two phases:

1. What is easiest to keep looking decent with the least time and trouble?

2. What is easiest to repair, if there is serious damage not involving the entire floor?

A good glossy varnish is very easy to keep clean and gleaming if it is not scratched, and particularly if it is not scratched through to the wood. Most dirt just wipes up. This is also true of lacquer. Semigloss materials don't show scratches as much as glossies, of course. Minor scratching can be kept invisible with wax—if the scratches are not too deep. Also, wax on a new varnish floor provides a sort of lubrication which forestalls scratching. In-the-wood finishes don't show scratches less serious than those you might suffer through some such action as someone dragging a nail across the floor.

Shellac and penetrating finishes are leaders in touch-up. You can wear out the spots in front of a door, the edges of stair treads, and other heavy wear areas, then touch them up easily with either of these materials. It is much more difficult to do this with varnish, lacquer, and other on-the-surface finishes. Lacquer floor finishes can be touched up all right provided the problem is wear—not failure of adhesion.

REMOVING THE OLD FINISH. Unless you are starting from scratch with a brand-new, sanded-and-ready floor, the first problem you face in floor finishing is removal of the old finish. Some people use a paint remover for this job; others go after the old varnish with boiling hot water and a strong trisodium-phosphate solution. Neither of these methods works half as easily or as effectively as a good floor sander. In every community there are building-supply dealers or hardware merchants who rent heavy drum sanders used for stripping old floors. The job is no harder than mowing a good-sized lawn with a walking mower, although the direction of force is different; you hold back on the sander, letting it move slowly across the floor by its own power, rather than pushing it.

Your objective when you sand a floor is to remove *all* old finish and *all* scratches in the wood and *all* stained wood. To do this, use medium to coarse drum belts (the man you rent the machine from can supply you) for the first two times over. Then switch to fine paper to clean up and smooth up. For a really good job, you must end up with hand sanding. A good trick is to put a couple of thicknesses of old blanket or equivalent around a brick and use it as a sanding block. Its weight is enough so that you don't have to bear down; just propel it back and forth, with the grain.

When a floor is particularly dirty and scratched, or when the boards are cupped or dished or washboarded, the best method is to make two passes on the diagonal across the boards first in one direction, then in the other.

When you do old floors over, the best practice is to sand them back to the bare wood. Handling a heavy sander is not difficult, and you can rent one readily at large hardware stores.

Do not ever sand across the boards, for if you do, you'll merely follow any dishing or washboarding without leveling it.

After these diagonal passes, make one grainwise pass with the medium-coarse paper. Then shift to fine for as many grainwise passes as required to ready the surface for hand sanding.

Owing to the bulk and design of the heavy drum sanders used in floor finishing, it is impossible to come close to edges. You can rent a heavy-duty disk sander to clean up edges with. It is almost imperative to lift the shoe mold (the molding at the bottom edge of the baseboard) so you can sand

Since a heavy machine can't get close to the edges, you'll need an edger, too. In the corners there is always a place that must be hand-sanded. Be sure all old finish is removed, since many good modern finishes work best only on the bare wood.

143

For a good job, floors must be hand-sanded close to walls, under radiators, etc. Even across the middle of the floor area, machines cannot equal hand work, if you want the smoothest possible, easiest-to-clean floors.

Staining floors the color you wish is easy with any of several different kinds of stains. Use oil stains or sealers under varnishes or penetrating finishes. Use non-grain-raising or water stains under lacquers, spirit varnishes, shellac, and other topcoats likely to be incompatible with oils.

Rent an electric buffer-polisher for the final waxing of your floors. Nobody can hand-rub floor wax well enough to take full advantage of its beauty and durability.

right up to the baseboard. Then, when you have finished sanding, nail the shoe mold back; it will cover the narrow strip that the sander cannot reach.

In corners, you must always use a scraper or a hand sander, since the disk type leaves a little triangular area that no sander will reach.

After you have sanded completely, vacuum the floor very carefully, and you are ready for the new finish.

STAINING THE FLOOR. The first step in putting on the new finish is staining or coloring, if you decide to follow the trend toward woodhue floors. These are the common methods of coloring wood floors:

Pigmented wiping stains—as discussed in Chapter 9—are handled the same for floors as they are for any other wood. Give them double drying time if you topcoat them with a lacquer finish, or you may run into bad adhesion. Drying is particularly important, since you most surely will leave a surplus of stain in the cracks. If this does not harden thoroughly, it will react with the lacquer and the result is almost certain peeling and flaking along the cracks.

Several standard colors in pigmented wiping stains, as made by a variety of manufacturers, are particularly good for floors. Generally, these are in the walnut or brown mahogany classifications. Get hold of a piece of flooring of the same wood as your floor and test some of these colors in half-pint sizes before you invest in a gallon or more for the whole job.

Colored penetrating sealers. Some lines of floor seals are manufactured in a range of excellent colors. Or, you can add colors in oil to regular clear sealers. See Chapter 9 for the technology involved in mixing your own stains.

Stains. You can use regular non-grain-raising or water stains. These will give you the best, clearest, and most permanent colors without any sealing action on the wood. They are the best under lacquer topcoatings, and produce superb colors under penetrating finishes, shellacs, or varnishes of any kind.

Colored varnish. If the color you like best happens to come from a colored varnish, experiment on a sample of the wood until you find how much dilution the color will stand, using turpentine or paint thinner. Thinned, the varnish will penetrate and carry much of its color into the wood. Then, topcoat with a good varnish system. Don't use lacquer over the colored varnish unless you give the varnish two or three full days of good drying weather.

All of these systems of coloring except water stains and NGR stains represent a degree of finish in addition to the coloration. The pigmented wiping stain usually carries with it some resins. The colored varnish has some resin content. The colored penetrating finishes are, of course, complete in themselves as one-coat systems. The extent to which these materials seal or build modifies the amount of topcoating required. However, most experienced finishers ignore all except the penetrating finish and proceed as though the wood were bare.

WHEN TO USE WOOD FILLER. Filler to smooth the pores of oak and other open-pore materials used for flooring is used much less often today than it was in the past. Generally, the only time you should consider it is when you want a mirror-smooth varnished floor. Filler used on floors is the same as you put on furniture (see Chapter 11). The only difference is the wholesale manner of application for floors. Thin the filler, tint it if required, and swab it on first with the grain, then across the grain. Then, when it has dulled over, wipe hard cross-grain with coarse rags such as old burlap bags. As with furniture filler, be sure to wipe clean. When the filler is dry, if you can feel a slight tooth when you rub your hand over it, chances are you didn't get it clean enough. A quick once-over with fine sandpaper, by hand, will take off the tooth.

Filler must dry hard before you topcoat it, particularly if you use lacquer over it. In some cases, you fill after the first coat of regular finish.

FINISHING SCHEDULES. The following schedules assume a sanded and dusted floor, and are based on the combined recommendations of flooring manufacturers and the finish makers.

Shellac. A 3-pound cut is recommended. Apply the first coat with full, smooth flow, using a wide brush. Avoid puddling. Allow this coat to dry two hours. Apply filler if required, and allow plenty of drying time. Hand sand with 2/0 paper and dust. Recoat. Allow the second coat to dry three hours, if you put on a third coat, four hours before you walk on it. Sand with 2/0 before a third coat. Sand with 3/0 for a final smoothing. Wax twenty-four hours later. (Filler is rarely used with shellac, since the shellac fills the grain quite well in two- or three-coat jobs.)

Lacquer. Use no filler. For the first coat, a special lacquer primer is suggested as optional for oak but required for pine and maple. Apply the first coat with a wide brush or a mohair roller. Work fast to prevent the double-build from area to area from producing lapmarks. Allow one hour for drying. Hand sand with 2/0, dust. Recoat. For two-coat work, allow overnight drying. For three coats, sand again with 2/0 and put on a third coat reduced with one part of thinner to four parts of lacquer. To avoid drying problems, use the thinner manufactured by the firm which made the lacquer. Allow overnight drying before traffic. Wax this finish when wear starts to make it dull.

Varnish. Thin the first coat of varnish with 1 part thinner to 8 parts of varnish except over penetrating wiping stain or colored varnish used as a first coat. Apply succeeding coats at full strength. Flow the varnish on as smoothly as possible, laying on, brushing out, and tipping off each brush full. Avoid puddling. Look at the varnish from a low angle now and then, to make sure you are not skipping or puddling. Allow each coat of varnish twenty-four hours to dry. Sand with 2/0 between coats. Dust carefully, and wipe with a turpentine-dampened rag between coats. Wax when wear starts to dull the surface—or immediately if you wish, to delay wear. *Urethane varnishes:* use semigloss or satin for final coat, to produce nongloss surface. Read labels for drying times, which are much faster with some urethanes than with regular varnishes.

Vinyl varnish. Seal the bare wood with 3-pound shellac. Let this seal dry two hours. Hand sand with 2/0 paper and dust. Apply vinyl varnish in smooth thin coat. Recoat *after* two hours but *before* six hours. If you cannot recoat before six hours, wait forty-eight hours. Waxing is optional. (Check labels carefully, since these materials often require special handling techniques.)

Penetrating wood finish. Brush, roll, or swab on a full, wet coat. Allow twenty to thirty minutes for penetration. Watch the floor during this period. If any spots turn dull it is a sign that they have absorbed all the sealer. Swab them with more. At the end of half an hour (some brands ask for more) wipe the excess off. For a better job, wait three hours (some brands ask for more) then repeat the application. Wipe as dry as possible. Some people wax this finish, although it is not necessary.

Flooring contractors often buff the sealer into the wood, using No. 2 steel wool on a machine. You can, if you wish, emulate their process by buffing the sealer with pads of steel wool, hand held.

CHAPTER TWENTY

THE PAINT BRUSH

A PAINT brush is the most important tool in the entire wood finishing process. It is true that some finishes require only the crudest sort of swab-on and could be applied with a worn-out broom. But any sort of decency in a shellac, lacquer, varnish, or enameled finish calls for a first-class brush.

It is difficult to understand why a person who has in mind a good wood finish will attempt to do the job with a bad paint brush. Many times, no doubt, the philosophy is, "I hate to clean paint brushes, so I'm going to throw the brush away as soon as I get done, and it doesn't make sense to spend a lot of money just to throw it away." For those of relatively unlimited wealth, this might be all right, except for the basic truth that you cannot do a good job with a bad paint brush.

WHAT A GOOD BRUSH DOES WELL. A good brush flows the finish on smoothly, and helps you spread it in an even coating. A good brush has bristles that taper to ends that are almost invisible—and these ends are "fibrillated", so the exact tips are almost infinitely small. As a result, they let the paint flow off the bristles without ridges or striations. While it is true that well-flowed finish will "level" itself, the less roughness the brush leaves, the less you must depend on leveling. As a consequence, you don't find yourself putting on material too heavily, as though troweling it on, in order to provide enough actual bulk to insure leveling. Moreover, you don't spend time trying to smooth out roughness with the brush that made the roughness in the first place—a sort of futile occupation.

A good paint brush helps you work faster. It carries more finish in its long, well-shaped bristles, so that you dip less often. Then, by laying the material on more smoothly, it saves you brush-back time.

Finally, a good brush has its bristles fastened in properly. They don't fall out and mess up the job. In many cases, the handle has a hole to hang the brush up by when it's not in use. Regardless of style or shape, which are matters of personal taste, the brush will feel good in your hand.

The ferrule should have a tight fit on the handle, not loose. It should

Whether the bristles are natural (left) or nylon (right), they should have tips which taper almost to invisibility, as those shown here do. This lets the paint off the bristles in a smooth layer, without the furrowed, rough look produced by blunt bristles.

be well fastened with at least two pins to a side—sometimes four or more. Most ferrules are plated for good looks and to prevent undue rusting when you clean the brush in water. In some cases, you'll find stainless steel ferrules; these are usually used only on high grade brushes—mostly when they are intended for use in water-thinned materials.

BRISTLES. Paint brush bristles are made of animal hair—or of synthetic fibers. The natural hair used is hog bristle and sometimes such exotic materials as fitch, camel, sable, or skunk. At one time, all good paint brushes of a utilitarian nature were made of pure China hog bristle—the guard hairs that grew several inches long through the regular coat of a tough wild hog. The supply of these bristles—usually pulled by the handful out of untamed and uncooperative wild boars—began to diminish in the thirties, and serious experimentation began among American manufacturers to find a substitute. Nylon turned out to be the reigning material, and today they make nylon bristles that are exactly as good as—or better than—the best natural bristles a hog ever grew. They do this by making a nylon bristle that you can barely tell from an animal bristle—except that it is shinier. They taper, as hog bristles do. They are mechanically fibrillated, as hog bristles are because of years of rubbing against the brush and rocks of the wilds. They flex, as the best hog bristles do. Most importantly, they lay on paint as smoothly and fully as hog bristles do.

What this means is that you don't buy a paint brush on the basis of what is printed on the handle; you judge the brush on the basis of bristle quality. The quick-to-see characteristics, bristle or nylon, are these:

The bristles must be long enough to flex well. A quick rule of thumb is that bristles should be about 50 per cent longer than the width of the

Good brushes have bristles about half again as long as the brush is wide. This provides good flexing, as demonstrated here. When you buy a brush, bend its bristles in your fingers. They should flex like a fishing rod —more at the tips than at the base.

brush. For example, a grade A brush two inches wide will have bristles about three inches long. At the extremes, this rule does not hold. A half-inch brush with bristles only three-quarters of an inch long would be stubby. (Beware, however, of the narrow brush with bristles so long that it is floppy.) At the other end of the scale, a four-inch brush doesn't demand a six-inch bristle. At about $35 or more, you could once buy a hog-bristle wall brush with six-inch or longer bristles, and this made you a painting magnate. Today the brush manufacturers' association only demands a six-inch bristle for a huge five-inch brush to be labeled AA. (See chart)

Typical Width, Length, Thickness of Quality Brushes

Width	Grade of Brush					
	AA		A		B	
	length	thick	length	thick	length	thick
1 in.					$2\frac{3}{16}$	$\frac{7}{16}$
1½ in.			$2\frac{11}{16}$	$\frac{11}{16}$	$2\frac{3}{16}$	$\frac{1}{2}$
2 in.			$2\frac{15}{16}$	$\frac{3}{4}$	$2\frac{7}{16}$	$\frac{9}{16}$
2½ in.			$3\frac{3}{16}$	$\frac{13}{16}$	$2\frac{11}{16}$	$\frac{5}{8}$
3 in.	$3\frac{13}{16}$	1	$3\frac{7}{16}$	$\frac{13}{16}$	$2\frac{15}{16}$	$1\frac{7}{8}$

Since it is the bristle tips which apply the finish, look for the tapered ends, shown in the accompanying photo. Some cheap brushes are made of the *butts* of the bristles. At some point between the hog and the brush, the bristles are cut in two. The good ends go into good brushes. The blunt-end butts go into cheap brushes. The same dodge can be employed with nylon bristles. At less cost, they put square-cut nylon fibers into a brush. You couldn't pick a worse tool for applying a finish. Look at the bristle tips. Animal or synthetic, they must not have blunt ends.

Some natural bristles are bleached to a golden color. The process tends to soften them somewhat. They are generally used only on fairly small brushes. The quality is fair, but usually not as good as good black bristles. Also, some imported nylon bristles are gold in color. In brushes of good manufacturers and at quality prices, these gold nylon brushes are as good as you can buy, excellently flexible, with good tips.

Not only must bristles be long enough and of good quality, but there must be enough of them. Superficially, you can judge this by observing the thickness of the brush. For example, a good one-and-a-half-inch brush has a bristle bundle about an inch thick. A three-inch brush will be a little over an inch thick. As with bristle length, there are extremes. A very wide brush is thinner, proportionately. It doesn't maintain the three-to-one ratio; it doesn't need to. On the other hand, a one-inch brush or smaller is usually thicker, proportionately. Otherwise, it wouldn't have enough thickness to handle properly.

Thickness alone, however is not the entire story. A brush may be thick without having enough bristles. As one of the photos shows, most paint brushes are made with one or more wooden—or aluminum or plastic— wedges which help hold the bristles firmly in the ferrule, and space the bristles properly at their butts. It is possible to skimp on bristles by using bigger wedges—or by using more of them than necessary. You can usually tell when a brush is under-bristled by squeezing it between your fingers at a point about in the middle of the bristles. It should have a good bulk at this point, and it should not seem to grow suddenly skimpy a short distance below the ferrule. When a brush has too few bristles it doesn't flex properly for smooth finish application and it doesn't hold enough paint.

A large brush cut in half shows how bristles are held between wedges in the handle. The wedges give the bristles proper spacing in the ferrule, since you don't need solid bristles in a big brush. Too many wedges, or too thick, may mean not enough bristles, however. The text tells you how to check on this.

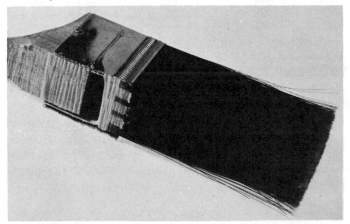

You dip constantly—and it drips constantly. (Although paint is not under discussion here, it is of importance to note that for *latex paints that are quite thick,* a shorter bristle and fewer bristles are considered best.)

Finally, a good brush is constructed with a wedge-shaped tip. As you look at the edge of the brush, you can see that the center bristles are longest, and they are increasingly shorter on the flat sides. This is called "chisel". Its purpose is to allow as many as possible of the *bristle tips* to be in contact with the surface as you stroke in either direction.

THE SHAPES OF PAINT BRUSH HANDLES. The shape of the handle has nothing to do with the quality of the brush—except for very cheap products with handles that are plainly die-cut from flat stock. You pick a handle for the way it feels. Some like them short, some long. Some round, some shaped. A flat handle feels best to many finishers. Others like a Kaiser. For a great many experts in wood finishing, a good brush intended for painting windows (called a "sash tool" in the trade) makes a perfect brush for working on the many shapes and areas of furniture.

Pick a handle shape you like, one that feels right in your hand. The brush at left is commonly called "Kaiser." At right is a beavertail. The two in the middle are flat-handle and round-handle sash brushes, basically intended for painting house trim and windows, but considered excellent for furniture and other finishing by many experts.

HOW TO TAKE CARE OF A PAINT BRUSH. The worst day a paint brush ever has is the first day you use it. After that, it gets better all the time—if you take care of it. This is what makes the purchase of good brushes a wise move, and proper care a most intelligent move.

Varnish and enamel brushes. It is not hard to clean brushes that have been used in oleo-resinous materials such as varnish and enamel. Just haul off and clean 'em. Use a cheap solvent to start with. Gasoline or kerosene is economical enough so you can use plenty. Treat the brush to two or three or more rinses in these cheap materials. When you are sure the varnish has been cut loose and flushed out, give the brush a final rinse in turpentine. Turpentine dries and leaves an almost undetectable protective film on the bristles. They stay soft.

Another good way to clean varnish or enamel brushes (or those used in any other finishing material) is with special brush cleaning formulas paint dealers sell by the pint, quart, or gallon. These cleaners contain several solvents, one of which is sure to be specific for the kind of finish involved. They are not cheap to buy, compared to gasoline, but they are often re-usable. The resins and pigments sink to the bottom, leaving the liquid clear, although sometimes slightly colored.

It is extremely sound practice to wash any paint brush in warm water and a detergent after you clean it initially in any solvent. You can use fairly warm water on nylon bristles, although natural bristles may curl up if you let the water get too hot.

Hang the brush up to dry. (If any paint brush you have doesn't have a hanging hole in the handle, drill one in it before you use it again; more brushes come with holes these days, and all brushes ought to.)

When the brush is dry, you can take your pick between two procedures . . . Either wrap it carefully in kraft paper, taking pains not to bend the bristles out of shape; or, hang it up bristles down inside a cabinet some place where it will be protected from dust and dirt.

Lacquer brushes. Use a lacquer thinner. If you work much with brushing lacquers, clear or colored, you may want to buy a cheap thinner in gallon quantities to use as a cleaner. The commercial brush cleaners work on lacquer, too. Follow the same water-wash methods as outlined above for varnish and enamel brushes.

Shellac brushes. Ammonia in warm water quickly cuts shellac and gives you a clean-as-new brush. Follow with detergent and water if you like.

Paint stores sell a variety of powdered brush cleaners which are quite effective, particularly if neglect has let finish harden or semi-harden. These cleaners tend to take a little of the life out of natural bristles. It is best to use them promptly after you are done with the finishing job, when they work at relatively weak concentrations, rather than to let a brush harden, then attempt to resuscitate it.

Brush storage tricks. If you are involved in a lengthy finishing project—or if you are in the business more or less continuously—there is little point in cleaning a brush between coats or even between projects. Professionals provide themselves with containers of "brush keeper," composed of raw linseed oil (two parts) and turpentine, (one part.) By some means, the brushes are suspended in this solution so that the bristle tips are at least an inch from the bottom. A good way to do this is to drill holes in all *brush handles an equal distance from the bristle ends.* That way, when the brushes are hung on a wire across the top of the container, all bristles are submerged to an equal distance, well up from the bottom. If you intend to store brushes in this sort of a container for a period of time, you must devise some sort of a cover to hold down evaporation. Some paint stores sell cans designed specifically for brush storage.

If you are a now-and-then finisher, concerned mainly with keeping brushes between coats of varnish or enamel, you need only store the brushes in turpentine. Don't use paint thinners which tend to cause a coagulation of the resins. Good turpentine keeps the brush soft and clean. When you are ready for the next coat, shake out the excess turpentine and wipe the brush as dry as possible on lint-free rags.

Be careful not to let the bristles rest on the bottom of the container, or they may take a permanent bend. To avoid this, hang them by a coat hanger wire hook over the edge of the can. Or, wrap the bristles and ferrule in several turns of brown paper, fold it up a half-inch beyond the bristle tips, and let the brush rest on the folded paper.

Naturally, you should not store brushes used for enamel and those for varnish in the same solvent, or you may transmit colors to the varnish brush. Do not, however, believe the legend that you must never use an enamel brush for varnish. If you clean a good enamel brush well after each use, it benefits from the work in enamel, and is also a good varnish brush.

There are brush-preserving jellies such as Stat, available at good paint stores, which coat the brush and prevent hardening of the finish on it. When ready to use the brush again, you just wipe off the jelly and proceed as usual.

A WORD ABOUT SPRAYING. Since a spray gun is a fairly expensive piece of equipment, along with its compressor, the manufacturers can afford to provide you with good instruction books—and they do. The manufacturers put out elaborate literature on the use of their guns, containing more useful information than there would be room for in a book on the entire finishing process. It is important, nevertheless, to make these points:

Brushing is scarcely able to match spraying for smoothness of application.

Spraying is considerably faster than brushing, and this is particularly true when you work on a project which has many spindles and shapes. You can spray three or four Boston rockers, for example, while the fastest brush in the neighborhood is painting one.

The blast of air from the bleeder type of spray gun (which is the best for average home use) is a good duster, blowing dust and dirt away before you spray the surface.

You can clean a spray gun more easily and quickly than you can clean a paint brush, provided you use enough kerosene or gasoline. However, if you let a spray gun dry hard, cleaning it becomes quite a chore.

GLOSSARY OF WOOD FINISHING TERMS

Abrasive. Any of the coated papers or fabrics or other materials including pumice and rottenstone and steel wool, used for smoothing wood or between-coat smoothing of finishes.

Adhesion. Ability of a topcoat to stick to a substrate or a previous coating.

Alkyd. A synthetic resin developed from various oils, used in varnishes and enamels as the film-forming material.

Alligatoring. Deterioration of a finish in the form of network cracks, caused largely by a difference in flexibility between a topcoating and the coat beneath it.

Aniline dye. A true dye material, soluble in form, made from oil derivatives.

Antiquing. Any of several processes used to impart the look of age to a finish or the wood beneath the finish. It includes "distressing" the wood, as well as glazing an applied finish with semiopaque antiquing liquids.

Base coat. The first coat applied in a finishing schedule—usually not including sealers, sanding sealers, or lighteners. Also, the base *color* of a glazing job.

Bleach. Liquid powder to be dissolved in water, used to lighten the color of wood. May be single-solution, for mild effect or a two-solution formula for almost complete bleaching.

Bleeding. Tendency of some stains and some bare woods to release color into a finish applied over them. Controllable by sealers.

Body. In general, the thickness or viscosity of a finishing material.

Boiled oil. Linseed oil which will dry. Formerly it was boiled to make it a drying oil. Today it has chemical driers added.

Bridging. Ability of a finish to fill over minute pores and cracks in wood or a previous finish.

Brightwork. Name given to varnished or clear wood finishes as opposed to painted. Usually used in marine finishing, but also in coachwork.

Brushing lacquer. Lacquer formulated with slow-drying solvents to allow time for brushing without showing brush marks and laps.

Catalytic coatings. Finishing materials which depend on a "hardener" for setting up and curing, rather than merely evaporation of thinners. Usually specialist in nature, high performers for certain finishing requirements.

Checking. Roughening of wood in the form of grain-wise cracks, usually caused by moisture. Occurs under finishes which may be waterproof, if moisture can gain access from behind. Term is also applied to grain-wise cracks in a finish which may not actually occur in the wood.

Close-grain wood. Wood which does not reveal open pores when it is dry. Examples are maple, cherry, birch, holly.

Craze. Condition of a finish similar to alligatoring except that the network of cracks is much finer.

Crocus cloth. An extremely fine abrasive used at the final stages of rubbing a finish, usually with a rubbing oil.

Dipping. A method of finishing small articles by dipping them in a container of the finish.

Drier. Formula available at paint stores by which drying times of some finishes can be speeded up. However, most finishes already have as much drier in them as they will stand, and the addition of more may be deleterious.

Dry colors. Dyes in powder form, dissolved in water, alcohol, or mineral spirits to form stains.

Enamel. Name given to colored finishes with a high varnish content, similar to varnish in handling and protection. Also used to designate colored lacquers.

Epoxy. Extremely tough and durable synthetic resin used in some of the catalytic coatings. Also formulated as a two-part adhesive.

Fat edge. Unsightly and deterioration-prone over-thick layer of a finish at any edge, caused by allowing the brush to flip or fall over the edge.

Feathering. Blending the edges of a finished area by lifting the brush at the end of the stroke, so that the edge becomes indefinite. Also, sanding an old finish around a blemish, so as to blend the finish film smoothly into the wood.

Felt. Used in the form of a block or applied in sheet form to wooden block to apply rubbing powders with oil. Felt over wood makes the best sanding block, also.

Filler. Pastelike material made of oils, driers, and ground silica, used to fill wood pores in species of open grain, which the finish alone would not bridge.

Firply. Familiar term for fir plywood. Used more extensively in Canada than in the U.S.

Flat finish. A finish without sheen, as the result of rubbing or because of flatting oils and pigments in the formulation.

French polish. Very lovely finish made by rubbing a mixture of shellac and oil on the surface. See Chapter 12. Finish is not very durable in hard-use situations.

Gesso. Pronounced jesso. A finishing technique utilizing plaster of paris or other form-taking material to provide a textured or formed base, over which a color or colors are applied.

Glazing. The process by which a second, semiopaque color is brushed on then wiped off selectively, revealing a base color. The process is often called "antiquing".

Glossy finish. Glass-smooth surface of enamels and varnishes which are not formulated with de-glossing materials.

Grain. The natural pattern of sawn and smoothed wood.

Graining. Simulation of wood grain, using a base coat and a glazing liquid. One of the glazing or antiquing tricks.

Hardwood. Name given generally to wood from deciduous trees as opposed to coniferous trees. Thus, pine is a soft wood; maple is a hardwood. (It is of interest to note, however, that larch is a softwood and basswood is a hardwood —but basswood is much softer than larch.)

Lacquer. Clear or colored finishing materials usually based on cellulose nitrate. Looseness of application, however, allows the word to be used to cover quick-drying materials based on other film-forming materials, such as acrylics and vinyls.

Latex. Genergic term used to cover a variety of finishes and paints which use water as the thinner.

Mineral spirits. Paint thinners and brush cleaners made as derivatives of oil.

Multisolvent. Cleaners for brushes or surfaces to be painted, formulated of several different solvent materials, each specific for a certain kind of contamination, so that the cleaners are universal in effectiveness.

NGR stain. The initials stand for "non-grain-raising". The product is a stain containing no water or so little water that it doesn't cause the swelling of fibers which raises the grain. Thus, no sanding is required after staining.

Oil stain. Stains formed by mixing oil-soluble dyes in an oil or an oleoresinous base. The term is sometimes applied to pigmented wiping stains, which may also contain pigments in suspension.

Oil-modified urethane. Modification of urethane varnish formulas which provides natural air drying, without hardener additives.

Open-grain wood. Wood which reveals conspicuous pores when it is dry. Examples: mahogany, walnut, oak, chestnut.

Orange peel. Faulty condition in which finished surface becomes broken or pebbled, most often result of incorrect spraying of lacquer or enamel.

Orange shellac. Unbleached shellac, which is a deep amber color.

Padding stain. Special powders which dissolved quickly in alcohol (i.e., shellac) or in special "padding lacquers", and are used for staining and blending with a pad of cloth for application. Used mainly for touch-up and for "aging" the look of under-age antiques.

Paste filler. Ordinary wood filler, in paste form, which must be reduced for application.

Penetrating finish. Any finish formulated of penetrating oils and resins which sink into the wood, leaving little or no material on the surface.

Phenolic. Excellent varnish resin made from phenol and formaldehyde. Used also in penetrating resin finishes.

Piano finish. The highest finish of all, produced with lacquer or varnish, rubbed and polished.

Pigment. Solid, fine-ground particles of color which do not go into solution but must be stirred into suspension in the vehicle. Compare with *dyes,* which are dissolved in the liquid.

Pigmented wiping stain. Stain formed of oils and resins in which pigments are suspended. In use, it is brushed on, then wiped off, leaving coloration of the pigments, plus sometimes soluble dyes as well.

Plastic finish. Loose generic term given to varnishes and lacquers of recent formulations not based on traditional ingredients. Includes urethanes, vinyls, acrylics, etc.

Polyurethane. Chemical term applied to oil-modified urethane varnishes which do not require hardeners or moisture-curing, but dry through evaporation.

Primer. Special material formulated for good adhesion and hold-out on wood and for good adhesion, in turn, of top-coatings. Used with enamels and paints, rarely with varnish or lacquer.

Pull out. The act of inserting bristle tips into a crack or a corner and drawing out excess finishing material.

Pumice. Powdered pumice stone, used with oil and felt for smoothing finishes. Pumice stone is solidified lava foam.

Reducer. A thinner, used for reducing the viscosity—thinning—varnish, shellac, lacquer, enamel, or other finishing materials. Each material has its specific thinner or class of thinners.

Resin. The material which forms the film of the finish. Formerly it was *rosin*, but today rosin means the solids left after the distillation of pine pitch, and resin means acrylic, alkyd, phenolic, or other manmade film formers.

Rosin. Solids left after pine pitch distillation. Used today only for making picking sticks.

Rottenstone. Very fine mineral powder, used for polishing.

Rubbed finish. Finish of varnish, lacquer, or enamel which is given infinite smoothness and a low luster by careful rubbing with pumice or with extremely fine waterproof abrasive paper.

Rubbing varnish. Formerly a varnish of short-oil formula specifically intended for rubbing. Today, any good floor or spar or glossy varnish which will take rubbing.

Sanding sealer. A special formula—usually a lacquer—or dilute shellac applied thin to the bare wood to make smoother final sanding possible.

Sap streak. Pockets of pure pitch often found in coniferous trees, frequently exposed in sanding and planing. The pitch must be sealed off (usually with shellac) since oleoresinous finishes will not harden over pitch deposits.

Scarify. To roughen a coat of finish so that the following coat will bond better to it. Required by the inability of most finishing materials to adhere to glossy surfaces.

Scuff sand. Basically the same as scarify. Indicates sanding of a coating only to a degree sufficient to provide a good adhesion.

Sealer. An initial coat on bare wood or over stain or filler, intended to prevent bleeding of undercoats into overcoats, or equally important to prevent penetration of finishing materials when the process is on-the-surface.

Semiglossy finish. A varnish or lacquer or enamel containing de-glossing chemicals which break up the surface shine. Usually considered to be a little more shiny than satin-finish, in a progression of gloss-to-flat which goes glossy, semiglossy, satin, flat.

Set. The initial hardening of a finishing material, sometimes called "dustfree", after which settling dust will not cling. Finishes harden considerably after the initial set.

Sheen. The reflectivity of a finish when viewed from a

fairly flat angle. Relates to glossy, semiglossy, satinsheen, and flat.

Shellac. Good, quick-drying finish for general use. Economical, easy to use. Not durable in wet or chemical situations, but sufficiently abrasion resistant when waxed to make a good floor finish. See also *orange shellac, white shellac.*

Softwood. Woods of the coniferous group. (See hardwood.) Also, the softer areas of woods which are notably less dense in some areas than in others, such as fir plywood and many other rotary cut veneers.

Spar varnish. Name given to varnishes intended for outdoor use. Most often, phenolic-resin varnishes. The term has little meaning, except that materials labeled "spar" are usually of good quality.

Spray. To apply a finish by means of spray guns or aerosol spray cans.

Spray can. Any of many popular aerosol products of lacquer, enamel, or shellac.

Spraygun. The working end of a spray system. Various types are available, the most useful of which for average use is the bleeder type.

Stain. Any material used to change the natural color of wood.

Stick shellac. Solid thermoplastic material available in most wood colors, applied in surface preparation, before finishing.

Substrate. Any surface over which a finish is applied.

Tack rag. A piece of fabric impregnated with varnish, oil, and water so that it remains permanently sticky under proper storage, and can be used to wipe off minute dust particles from a surface to be finished.

Thinner. Any material used to reduce a finishing material, i.e., reduce its viscosity.

Tipping off. The brushing technique which involves the use of the tips of the bristles to smooth the applied finish.

Topcoat. Any finishing material used *over* another. Purely, the term means the final coat, but in practice, the "topcoat" is any material applied over a "substrate".

Turpentine. The distillate of the pine-pitch system. Used to thin all oleoresinous finishing materials.

Undercoater. A finishing material with strong adhesive characteristics to wood, and the ability to form a film of strong adhesive characteristics to topcoatings.

Urethane. An important variety of resins for the varnish and enamel industry. Formulated both for catalytic (separate hardener) uses and for normal air-drying. An intermediate form is moisture curing, forming hard surfaces of extreme value for heavy-abuse situations.

Varnish. The oleoresinous clear finishes. Their films form upon the evaporation of thinners or vehicles, leaving resins which usually harden further, either by oxidation or chemical curing. The hardening characteristic of varnish is contrasted with the "remainder" or "residue" film forming of lacquers, except that the newer lacquers benefit from curing or hardening times, as varnishes do.

Varnish stain. Varnish in which oil-soluble dyes are dissolved. In use, these materials provide a coloration along with a protection. Generally, the color is not as good as you'd get from stain, and the protection is not as good as you'd get from varnish.

Vinyl. One of the synthetic resins, used to make paints often formulated with xylol or xylene so that quick-dry is a useful characteristic, although complete cure may take some time. Vinyl finishes are used mainly as opaque finishes—latexes—but are also put up as clears.

Water stain. An aniline dye stain made by dissolving soluble dyes in water. In general, water stains are the best colors and the most durable you can get.

White shellac. The bleached, creamy-white form of shellac, more often used than orange shellac.

Wood filler. The same as paste filler. A material used to fill the open pores—grain—of the wood, so that subsequent finishes are entirely level.

INDEX